GOING GREEN
WITH THE
INTERNATIONAL RESIDENTIAL CODE

Scott Caufield

Australia • Brazil • Japan • Korea • Mexico • Singapore • Spain • United Kingdom • United States

DELMAR
CENGAGE Learning

Going Green with the International Residential Code
Scott D. Caufield

Vice President, Technology and Trades
    Professional Business Unit:
    Gregory L. Clayton

Product Development Manager: Ed Francis

Director of Building Trades:
    Taryn Zlatin McKenzie

Development: Dawn Jacobson

Director of Marketing: Beth A. Lutz

Marketing Manager: Marissa Maiella

Production Director: Carolyn Miller

Production Manager: Andrew Crouth

Senior Content Project Manager:
    Andrea Majot

Art Director: Benjamin Gleeksman

For product information and technology assistance, contact us at
**Professional Group Cengage Learning Customer &
Sales Support, 1-800-354-9706**

For permission to use material from this text or product,
submit all requests online at **cengage.com/permissions**.
Further permissions questions can be e-mailed to
permissionrequest@cengage.com.

Library of Congress Control Number: 2010921848

ISBN-13: 978-1-4354-9729-0
ISBN-10: 1-4354-9729-5

Delmar
5 Maxwell Drive
Clifton Park, NY 12065-2919
USA

Cengage Learning is a leading provider of customized learning solutions with office locations around the globe, including Singapore, the United Kingdom, Australia, Mexico, Brazil and Japan. Locate your local office at: **international.cengage.com/region**

Cengage Learning products are represented in Canada by Nelson Education, Ltd.
For your lifelong learning solutions, visit **delmar.cengage.com**
Visit our corporate website at **cengage.com.**

Printed in the U.S.A.
1 2 3 4 5 XX 12 11 10

ABOUT THE AUTHOR

Scott D. Caufield is the Building Codes Administrator for Clackamas County, Oregon, where he has worked for more than 18 years. There he has held a variety of technical, supervisory, and administrative positions all in the field of building code enforcement. For the last 7 years, Mr. Caufield has served as the county's Building Official. He also serves as an advisor to the Clackamas County Office of Sustainability and works closely with the Clackamas County Citizen Advisory Board on Sustainability.

Mr. Caufield holds numerous professional certifications within the state of Oregon and nationally through the International Code Council (ICC). His Oregon certifications include Building Official, Fire and Life Safety Plans Examiner, 1 & 2 Family (Residential) Plans Examiner, Structural Inspector, and Mechanical Inspector. Nationally, Mr. Caufield is certified as a Building Plans Examiner, a Residential Plans Examiner, a Residential Inspector, and in other areas as well. Additionally, Mr. Caufield is a Certified Building Official (CBO) through the ICC.

In 2003, Mr. Caufield was honored with an appointment to the Oregon Residential Structures Board (ORSB) by Governor Theodore R. Kulongoski and he is currently serving his third term as the Building Official member of that board. The ORSB is an advisory board that assists the Director of the Oregon Department of Consumer and Business Services in the administration of the Oregon Residential Specialty Code, Oregon's state-amended version of the International Residential Code.

Mr. Caufield served as the president of the Oregon Building Officials Association (OBOA) in 2008 and 2009 and is active in OBOA's educational and outreach activities. He has developed training and educational seminars in the fields of green building, communications, customer service, and continuity of operations planning (COOP); and he has delivered dozens of educational seminars through his involvement with OBOA and as President of Scott Caufield Enterprises, a consulting firm specializing in the development of training for the public sector and building code consultation.

Mr. Caufield's first interest in sustainability and green building—particularly in the area of energy conservation—dates back to 1993 when he earned the Certificate in Energy Conservation through OBOA's voluntary certification program, one of only 10 persons in Oregon to have obtained this distinction.

ACKNOWLEDGMENTS

As with any creative work, there are many people whose contributions make it all possible and this book was certainly no exception! I would like to acknowledge and thank the following persons for their contributions, without which this book would not have been possible.

To the entire team at Delmar Cengage Learning, thank you for your support. You have all been so gracious and have made me feel welcome as an author. I would especially like to thank Dawn Jacobson in Development for your encouragement, guidance, and patience. It has been a joy working with you on this book. To Edward Francis, Product Development Manager, thank you so much for believing in me and for giving me the opportunity to write the *Going Green* series; it is truly appreciated.

A special thank you goes out to the entire staff at the Clackamas County Building Codes Division; I am so very fortunate to know and work with all of you. You have all been amazingly supportive as I worked on this book and I am eternally grateful. To Cam Gilmour, Director of the Clackamas County Department of Transportation and Development, I offer my thanks for your support, flexibility, and understanding as I prepared this manuscript. You have consistently supported and defended the important work of the Building Codes Division and we are all indebted to you for it. To Eugene Morgan, Electrical Inspector Supervisor, and Wayne Seiffert, Plumbing Inspector Supervisor, thank you so much for your technical expertise and guidance. To Rod Jones, Electrical Inspector, thank you for the information on green electrical systems. To Kelly Sumetz, Plans Examiner, thank you for your interest in and passion for green building—I enjoyed our many conversations on the subject.

Thank you to the Oregon Building Officials Association (OBOA) for allowing the use of their Alternative Means and Methods Request Form, which has been adapted for use in this book. The OBOA is comprised of some of the most dedicated code professionals I have ever known and I am so very proud to be a part of such a fine organization.

A special thank you is due the International Code Council for allowing the use of the many tables and figures that appear in this text. Thank you also to the International Code Council-Evaluation Service for the use of the sample evaluation report, the sample SAVE-Verification of Attributes report, and the "How to Read" guide that appear in Appendix B. All tables and figures courtesy of the 2009 International Residential Code, Copyright 2009. Washington, D.C: International Code Council. Reproduced with permission. All Rights Reserved.

Thank you to my family, friends, and neighbors for all your support and encouragement. Thank you especially to my stepsons, Tracey and Scott, for doing all the household projects I should have been doing. Thank you to my daughters, Angela and Jennifer, for understanding when I could not be there for you as writing deadlines loomed. Finally and most important, I would like to thank my loving wife, Cindy, for her extraordinary support all these many months as I prepared this book for print. All of the good fortune and opportunities in our life together have been made possible because of you. Thank you for everything.

To those who reviewed the manuscript as it was being prepared, I am deeply grateful to you for your insight and expertise. This book is better because of you. The author and Delmar Cengage Learning wish to thank the following reviewers for their invaluable contributions during the development of this text.

Tom Phillips, Building Official
City of Salem, Oregon

David Eisenberg, Director
Development Center for Appropriate Technology

Steven Peterson, Associate Professor of Construction Management
Weber State University

John Schaufelberger, Professor; Chair
Department of Construction Management
University of Washington

Hamid Naderi, PE, CBO; Principal Staff Engineer
International Code Council

Kraig Stevenson, Staff Liaison
International Code Council

A special thank you is also due Mark Mecklem of Miranda Homes for the photos that appear on the cover of this book and in various locations throughout the text. Thank you also to Rob Boydstun of Miranda Homes for access to your construction sites and source material.

INTRODUCTION

This book was written with the hope that it would become an indispensable tool for those who seek to build a green dwelling. Some of the material presented, however, is fully applicable to *all* types of construction; thus the book's usefulness extends well beyond the green project. Many of the principles presented in the text are universal, meaning they can be just as easily applied to conventional construction as they can to green materials and construction methods. Builders, design professionals, and code officials will find this especially useful.

It may take more than one reading to fully grasp the scope of the material covered in the book—and that is okay. It often takes many years for code officials, plans examiners, and inspectors to become experts in the field of code enforcement, so the reader should not expect to become an expert overnight. The book will be most useful to those who invest the time needed to understand the concepts presented. The following sections establish the basic structure of the book, in an effort to assist in the understanding of its content.

How This Book Is Organized

For the reader to successfully integrate green building materials and methods with the codes, it is essential that he or she understand how the building codes function and what it is the codes are trying to do. Thus, Part I begins with a thorough but concisely written building code primer.

The primer will help the reader to understand the role of the code official, the reason for adopting codes, and the ways in which the building code generally regulates construction materials and methods. This introductory material will help the reader to establish a way of thinking about the building codes and, more importantly, how to link green building requirements with the various requirements of the code.

In the case of *Going Green with the IRC*, the primer assumes the reader has little if any experience with the building codes. Even if you have some experience with the codes, you are encouraged to read the entire primer as it may offer certain "revelations" that assist you in your quest toward building a green project.

Broadly, this book addresses two key areas in which the 2009 edition of the IRC establishes criteria: building materials and construction methods. The code establishes standards for building materials to ensure that products and materials used in construction are safe, durable, structurally sound, and, where appropriate, fire resistive. Additionally, the code prescribes construction methods to ensure that materials are used properly and to ensure that buildings are constructed using time-tested, safe building practices.

Because the choice of building materials is a key area in green building, the book will first address building materials and the way in which they are regulated in the code. Information is provided to help the reader understand the code requirements for building materials and then how to relate them to choices for green materials. Tools are provided to help the reader evaluate green building materials to ensure they are acceptable under the code.

Sometimes, going green means selecting an entire building system or alternative form of construction. For example, straw bale construction and green roofs are both increasing in popularity. These systems not only utilize green building materials but they also require special forms of construction that may or may not deviate significantly from the requirements of the code for "conventional" construction. Information is provided to help the reader understand how these systems are regulated and provides guidance to ensure that the green building system chosen could be accepted by the code official.

Part I closes with an explanation of the alternative means and methods (AMM) provisions of the code. These brief but important code provisions give the code official wide latitude when considering alternate forms of construction, as are commonly encountered in green construction. A thorough understanding of this important section is necessary because many of the building materials and construction methods/systems used in green building require approval under these provisions.

Part II provides a chapter-by-chapter analysis of the code, identifying key areas in which "green" decisions need to be made. These decisions are generally presented using a three-tiered approach, as follows:

1. Establish and explain, in plain language, the requirements of those sections of the code that are candidates for a green alternative
2. Identify the green decision to be made
3. Provide solutions that enable the reader to successfully carry out the decision including product evaluation criteria and additional resources

Part II is organized in roughly the same manner in which the IRC is laid out so that the reader can simply follow along with the requirements of the code in much the same way as a building is planned and constructed. For example, the topics of site preparation and demolition are presented first, followed by building planning, then foundations and retaining walls, floor framing, and so forth. This ensures a logical, easy-to-read format that works well alongside the IRC.

Finally, an Appendix is provided with examples of product evaluation reports, a sample Alternative Means and Methods Request Form that is suitable for adaptation and use by the reader or the jurisdiction, material devoted to the preservation of existing buildings, plus other additional resources.

PREFACE

This text is designed to provide an understanding of and the interrelationship between the International Residential Code (IRC) and the world of green construction. It is not meant to be an exhaustive text on green construction materials and methods, per se. At best, such a text would merely be a snapshot in time in the rapidly changing world of green building. It would also become quickly outdated, as new green building products and systems seem to emerge onto the market almost daily.

Rather, this text will provide an understanding of and a way of thinking about the IRC that will work regardless of the particular type of green materials or systems chosen and without regard for the particular U.S. region where the reader happens to reside—now and in the future. The book content can easily be updated with each new code cycle (currently every 3 years) and as new "green" technologies emerge in the marketplace. As a result, this text will, for the most part, enjoy a long life.

The International Residential Code is used currently in 48 states as well as in Washington, D.C., and the U.S. Virgin Islands. This code is used or referenced in other countries as well. Interest in this text, then, will not be limited to one particular region of the country or even solely within the United States. Widespread use of the IRC makes this text particularly relevant for anyone seeking a deeper understanding of the residential code and its relationship to green building. Thus, a very large potential market exists for the *Going Green with the IRC* title.

Written in a nontechnical, easy-to-understand format, this book will have broad appeal to homeowners, builders, design professionals, and even code officials across the United States, basically anyone who has an interest in the ever-expanding field of green building. Richly illustrated, this book provides visuals of many types of construction and also a number of helpful tools such as learning objectives and forms that can be used as templates to assist with your project.

How Will I Benefit from This Book?

Whether you are considering green construction from the perspective of a homeowner, a builder, a design professional, or code official, this text will provide valuable information to assist you in that quest.

Homeowners

As a homeowner (or prospective homeowner), you will find this book especially useful because it provides critical insight into the operation of a modern building department and it will help you better understand the International Residential Code, especially the way in which it regulates green construction. The text begins with a useful building code primer that provides you with everything you need to and *should* know about the building code and the people who enforce it.

This knowledge will help you avoid costly mistakes because you will know how to properly evaluate green construction materials—or any construction materials—before you buy them. It will help you to evaluate green construction methods and systems to determine what is necessary to ensure a code-compliant project. You will avoid unnecessary delays because you will know and understand what your local building department needs to process your permit application, *before* you make your application.

Going Green with the IRC will also help you to establish an effective partnership with the code official in your area, an absolute must for a successful construction project. Most important, it will increase your chances to successfully integrate green construction into your building project in a code-compliant way.

Contractors

In addition to the same benefits this book provides to homeowners, as a builder, you will also find this text extremely helpful (see the preceding section). Many builders today act as liaison between their local building department and their customers. In that role, the knowledge gained in this text will help you to communicate more effectively with your local building department and with your clients as well.

This increased knowledge will make you appear more professional to code officials because you will truly understand how the local building department operates, its duties and responsibilities, and, most important, how the IRC regulates construction. This, in turn, makes you appear more professional in the eyes of your clients because you will be able to guide them expertly through what is often viewed as a complex regulatory process.

Going Green with the IRC will help you establish a better working relationship with the code professionals with whom you will meet and work, because you will know and understand what it is they are seeking to accomplish. Additionally, you will be able to "speak the language" of the code official, which is essential in building a trusting partnership with your local jurisdiction.

Architects and Engineers

As a design professional, you will also enjoy this book and find many bits of useful information. On the whole, architects and engineers have a clear advantage in understanding building department processes because of their training, education, and daily practice; but even the most seasoned design professionals can have gaps in their knowledge as it relates to the enforcement of building codes.

This text will assist you in the specification of construction materials because it provides you with valuable insight into the standards requirements of the IRC for all building materials. This is especially true when it comes to green construction, where many of the materials often used have limited or no testing, to ensure that they will meet the requirements of the code.

Going Green with the IRC will also help you to build stronger relationships with code professionals in those jurisdictions where you practice because you will have a deeper understanding of the code and keen insight into the operation of a modern building department. The section on alternative means and methods will also prove useful, as it provides insight into a powerful and often overlooked tool.

Code Officials

As a code official, you will benefit from this book, despite your existing knowledge of the code, because like most code officials, you need further experience dealing with green construction. How much exposure have you had to anything other than conventional construction? This book provides an excellent overview of the many green alternatives available for use in construction today and useful tips for dealing with other than prescriptive construction.

Going Green with the IRC also provides you with tools that will assist citizens in your jurisdiction in understanding the residential code. The primer is especially useful in that regard. You may wish to suggest it as essential reading for all of your customers prior to making an application for permit, as it provides the "stuff you wish everybody knew" regarding the building code. Examples of ICC-ES Evaluation Reports are provided, as well as a thorough discussion of this topic.

If you are unfamiliar with "green" construction, then this text will prove to be an excellent resource for some of the more common or popular green construction materials and methods and the way they interface with the codes. In that regard, it will provide useful materials for you and your plan review and inspection staff, especially Part II of the text, which provides a chapter-by-chapter IRC analysis.

If you or your staff has limited experience working with alternative means and methods (AMM), this text will provide a way of thinking about and an approach for dealing with AMM that is useful not only for green construction methods but for other unusual or nonconventional forms of construction as well. Finally, Appendix C will prove useful to you as it provides a comprehensive, functional Alternative Means and Methods Request Form that can be adapted for use in your jurisdiction.

Competing Texts

There are, to my knowledge, no other publications currently in print that specifically address the interrelationship between green building and the international codes. Many publications available today address green building codes from a land use perspective; that is, they address the use of green building products and methods from a *planning and zoning* perspective. They caution the reader to check with local zoning codes and ordinances to address matters such as building height, setbacks, and the like; however, none address the important link between green building and the building codes.

While planning and zoning rules are an essential part of the permitting and construction process, they are but one element of the process. The building codes must also be considered and it is this area specifically that is most problematic for the average homeowner, contractor, or design professional. The content of *Going Green with the IRC* focuses on this important area.

Why I Wrote This Book

There are currently a number of texts on the market that deal with the topics of green materials, green construction methods, and green development. These texts address a wide variety of topics and offer thoughtful, in-depth analysis on all things "green." While the titles are many and the coverage exhaustive, few if any address the topic of building codes and the important relationship between the two.

There are also numerous texts available on topics related to the various building codes and, of course, there are the codes themselves. These texts focus on establishing standards for construction materials and prescribing safe construction methodologies. The codes themselves do not address green building. Although it is a commonly held belief that the codes actually discourage green building, in fact, they neither promote nor hinder it. Instead, the codes establish an understandably high bar to ensure that construction materials are safe for use.

Thus, a unique problem exists in that there are no texts that *specifically* relate green building materials and construction methods with the requirements of the building code. In my experience as a code official, it is this lack of connection between the two that can cause significant problems during the permitting process and also during construction. For example, a well-meaning homeowner or contractor might spend a considerable amount of time and money researching and acquiring the latest and greatest "green" building material only to discover during the permitting process or, worse, during construction that it cannot be used because it does not meet the standards established in the building code for construction materials. This causes frustration and unnecessary delays, often at considerable expense.

This book provides the necessary link between the two. In other words, it provides the tools that will enable the homeowner, builder, or design professional to understand the role of the code official, how the code establishes standards for construction, and how to integrate green building materials and systems into the construction project in a code-compliant way. It also provides tools for the code official to educate those seeking permits about the important relationship between green construction and the codes.

What This Book Is *Not*

The preceding sections establish pretty clearly what this book is meant to do. However, it is also important to understand what this book will *not* do. For example, this book is not meant to be a definitive resource on all matters "green." There are already a number of texts and reference materials on the market that deal with the latest and most innovative green technologies and materials. As you have probably already discovered, there is considerable debate over whose materials or methods are greenest and best suited to green construction.

In fact, in my experience there is little agreement on just what "green" is in the first place. Does "green" construction mean the use of materials that are harvested or manufactured in a sustainable way? Or, does it mean the use of construction methods that do not deplete natural resources or harm the environment? Does it mean energy-efficient construction? Does it mean the use of reclaimed/salvaged materials? All of the above? As you can see, the questions are numerous! Given that these questions and issues are not likely to be resolved any time soon, this book will make no attempt to settle the debate. Rather, I will leave it to the reader to determine what green material or construction method best suits his or her needs and satisfies those values that caused the move to *go green* in the first place.

Additionally, this book is not meant to be a substitute for the International Residential Code or any other code. While this text does discuss the IRC in a thorough, comprehensive way, it addresses only a small number of the actual provisions that appear in the code. All readers are encouraged to refer to the 2009 edition of the International Residential Code published by the International Code Council for a complete presentation of the code provisions, available through Delmar Cengage Learning, at www.delmarlearning.com. If you wish to explore and understand the code more thoroughly, there are a number of high-quality supplements and reference materials available as well. These products, such as the *2009 International Residential Code Study Companion*, the *Significant Changes to the International Residential Code—2009 Edition,* and other titles, are also available through Delmar Cengage Learning at www.delmarlearning.com.

About the International Code Council

The International Code Council (ICC) is a nonprofit membership association dedicated to protecting the health, safety, and welfare of people by creating better buildings and safer communities. The mission of ICC is to provide the highest quality codes, standards, products, and services for all concerned with the safety and performance of the built environment. ICC is the publisher of the family of the International Codes® (I-Codes®), a single set of comprehensive and coordinated model codes. This unified approach to building codes enhances safety, efficiency, and affordability in the construction of buildings. The Code Council is also dedicated to innovation, sustainability, and energy efficiency. Code Council subsidiary, ICC Evaluation Service, issues Evaluation Reports for innovative products and Reports of Sustainable Attributes Verification and Evaluation (SAVE).

**Headquarters: 500 New Jersey Avenue, NW, 6th Floor,
Washington, DC 20001-2070
District Offices: Birmingham, AL; Chicago, IL; Los Angeles, CA
1-888-422-7233
www.iccsafe.org**

Disclaimer

This book presents the opinion of the author alone. No endorsement of any particular manufacturer, construction material, or construction method is expressed or implied. Unless noted otherwise, no affiliation with any of the organizations mentioned is expressed or implied. The inclusion of the ICC name and logo in this book does not signify approval of the entire content.

Despite the fact that this book will improve your chances of successfully integrating green materials and methods into your construction project—in fact, greatly so if the provisions of the text are carefully followed—there is no guarantee that the code official in the jurisdiction in which you live will accept your proposal for green construction. Thus, there can be no guarantee that building permits will be issued simply because you have followed the suggestions made in this book.

Finally, while the IRC is used in almost every area of the country, it is important to understand that many states and local jurisdictions amend the IRC locally. It would be impossible for a text such as this one to address all the possible iterations and amendments that could be made on a local level. Readers are strongly encouraged to check with their local building department early to ensure that all local amendments, ordinances, and additional codes are known and understood.

Regards,
Scott Caufield, CBO

CONTENTS

PART **I**

Part I of this book is intended to provide the reader with an understanding of the function and duties of a modern building department, as well as to provide valuable insight into the residential building code that will increase the chances for a successful, code-compliant green building project. It may be used as a stand-alone section or in conjunction with Part II of this text. Review it as often as necessary to become comfortable with the people, terminology, and concepts typically encountered in a local building department. All references to the "IRC" or "the code" mean the 2009 edition of the International Residential Code, the scope of which covers detached one- and two-family dwellings and townhouses not more than three stories in height.

BUILDING CODES 101: A PRIMER

A friend of mine once quipped after a long and frustrating experience attempting to build his own home that the building code was "unknowable." He was particularly frustrated with the permitting process and the issues that emerged during the review of his plans for code compliance. It seems that every time he thought he had his arms around the code's various provisions, every time he thought he had covered all his bases and addressed all the issues, some new twist would emerge out of the code that caused delays or caused him to change something he planned to do during the construction.

It saddened me to see how frustrated he was with the permitting process and how disillusioned he had become with the building code and the people who enforce it. What should have been a helpful and rewarding experience turned out to be a stressful, anxiety-filled encounter that left him feeling angry over weeks of unnecessary delays and increased construction costs. Mostly, though, it saddened me to think he was convinced that the code could not be understood, because in reality it is inherently "knowable" if one is simply provided with a little guidance.

It is out of that encounter that the idea for Part I of this book and in particular, this chapter, was born. Over the course of my career, I have seen my friend's scenario play out time and time again. I wanted to provide some tools that would enable the average person to "know" the code and to understand the way in which a building department operates. It is my belief that a little knowledge and insight into the theories behind the code and the people who work in this field will help pave the way toward a smooth project and eliminate a great deal of the frustration and delay my friend experienced.

This is especially true for green construction materials and methods because many of these technologies are emerging faster than the code can be updated or keep pace. The residential building code is largely equipped to deal with conventional construction; however, green building materials and systems often fall well outside what is regarded as "conventional." Thus, there is an even greater potential for the frustration, expense, and delays experienced by my friend because *going green* often falls into the "gray" areas of the code.

Whether you are a homeowner, builder, or design professional, it is of paramount importance that you understand the International Residential Code (IRC) and the way in which it regulates construction. It is also essential that you understand and develop the relationship with your local code official. This cannot be overstated. A well-formed relationship will pay big dividends when you finally apply for building permits in your area. Knowledge of the building code and of the operation of a building department will increase your chances of success significantly. The following sections offer critical insight into the duties, functions, and operations of a building department and form the first part of your journey toward a successful green building project.

SO . . . WHY DO WE NEED A BUILDING CODE ANYWAY?

Properly enforced building codes are necessary to provide communities and the occupants of buildings with safe, comfortable surroundings. These codes ensure that buildings are fire safe, structurally sound, and energy efficient. They also ensure that building systems function as intended to provide heating and cooling for human comfort, to prevent disease, such as with properly installed plumbing systems, and to provide safe, dependable electrical wiring systems. Later, we will see how these codes specifically apply to green building. But first, let's explore each major area of the code more fully so that we can identify key terms and ensure that the main building code concepts are understood.

Fire and Life Safety

One of the primary functions of a residential building code is to ensure life safety for the occupants of a building. Properly enforced, these codes minimize the risk of fire (to the maximum extent feasible) and ensure that, should a fire start, the occupants of a dwelling will be notified as early as possible and that the building can be exited quickly and safely through the use of safety features such as working smoke alarms and code-compliant exits.

Residential building codes also provide basic requirements for other life safety items such as emergency escape and rescue openings, which allow for exiting directly out of a bedroom in a fire and which allow a firefighter access for rescue operations. These codes establish requirements for safety glazing in required areas and they also establish standards for **flame spread** (the rate at which flame burns along the surface of a material) and **smoke density** (the quantity and density of smoke produced when a material burns) for wall and ceiling finishes.

For example, the code establishes requirements for **smoke alarms** (also called smoke detectors) to be installed in all sleeping rooms, outside each sleeping area, and on each floor level. Working smoke alarms provide early detection of fire by sensing minute amounts of smoke (see Figure 1-1). They also provide early warning to the occupants of the dwelling by sounding an alarm. It is widely recognized that early detection and early warning greatly increase the chances that the occupants of a dwelling can exit safely during a fire. This is especially true when the occupants of the dwelling are sleeping, a time when they are most vulnerable.

The code also attempts to reduce the risk of **conflagration,** or the spread of a fire from one building to another. It does so by providing safe clearances from the exterior walls of structures to property lines and by placing restrictions on and even prohibiting openings such as windows and doors to reduce the chances that fire will spread from one building to another.

Why is this important? Well, let the lessons learned from The Great Chicago Fire of 1871 serve as a great reminder of just how devastating the spread of fire can be (see Figure 1-2). In this case, a whole city burned because buildings were placed

Figure 1-1

A smoke alarm installed in a dwelling. Working smoke alarms alert occupants to smoke at the earliest stages of a fire, which is especially important while sleeping.

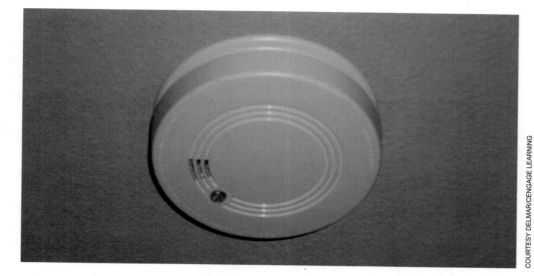

COURTESY DELMAR/CENGAGE LEARNING

Figure 1-2

The Rush for Life over the Randolph Street Bridge, 1871 (*Harper's Weekly*), from a sketch by John R. Chapin.

COURTESY OF CHICAGO HISTORY MUSEUM

close together and were made of materials that ignite and burn easily. No provisions were made in those days to construct buildings in such a way that, should a fire start, it would be stopped before spreading to another structure through the use of noncombustible materials such as brick or stone.

In cases where residential buildings are placed less than 5 feet (1,524 mm) from a property line, the code requires exterior walls to be **fire resistive**. That is, they must be constructed using wall construction methods that have been tested in a laboratory setting and proven to be resistant to fire for a period of at least 1 hour (see Figure 1-3). When constructed in this way, the passage of fire is restricted, which provides time for occupants to exit safely and for firefighters to respond to fight the fire. When exterior walls are less than 5 feet (1,524 mm) but at least 3 feet (914 mm) from a property line, the area of openings such as windows or doors is limited to 25% of the wall area. When exterior walls are less than 3 feet (914 mm) from the property line, openings are not allowed.

Figure 1-3

Two layers of fire-rated gypsum board at a common wall between dwelling units. Each additional layer of gypsum board increases the fire resistance of the wall.

You may have noticed an increase in the number of townhouses or rowhouses being built in your area. **Townhouses** (and similarly, rowhouses) are dwelling units built one next to the other, with no space between the units. Townhouses typically share one or two walls with an adjacent dwelling unit and have open yards on the remaining sides. Although these dwelling units are attached, the IRC regards them as separate buildings. In these types of structures, walls that are common to adjacent dwelling units are constructed on a real or an assumed property line. The residential building code establishes special criteria for these conditions to protect against conflagration. Here, fire-resistance requirements are increased and **penetrations** into these walls are closely regulated. Penetrations are construction elements such as pipes, ducts, and wires that enter or penetrate into or go through the fire-resistive wall. Again, the idea here is to prevent flame and hot gasses from spreading from one dwelling unit to another.

The life safety provisions of the code also establish standards for the construction of stairways, landings, and other exit components to ensure that a dwelling can be exited safely during an emergency. For example, the code establishes requirements for **guards** (also called guardrails), which are walls, railings, or other barriers to be provided around raised floors such as decks or balconies and at stair landings (see Figure 1-4). Guards must be constructed to a minimum height—36 inches (914 mm) in most cases—and must resist a concentrated load of 200 pounds (890 N) applied at the top of the rail. Openings in the guard (such as the open spaces between pickets) are restricted in size and must be constructed such that a 4-inch (102-mm) diameter sphere cannot pass through.

Structural Systems and Components

A second and no less important function of a residential building code is to ensure that buildings are structurally sound. Sound construction and properly designed structural systems are essential to ensure that buildings can withstand the effects of gravity, wind loading, and in some areas **seismic** (earthquake) events. Structural systems form, essentially, the skeleton of the structure and act much like the skeleton in your body, providing the support necessary to keep the building upright under a variety of loading conditions (see Figure 1-5).

Figure 1-4

A half-wall adjacent to a stair opening to be used as a guard.

COURTESY DELMAR/CENGAGE LEARNING

Figure 1-5

General wall framing—the "skeleton" of the structure.

COURTESY DELMAR/CENGAGE LEARNING

Gravity, an ever-present force, acts on buildings in a downward direction, toward the earth. That is, it has the effect of pulling roofs, ceilings, floors, and walls toward the ground just like it has the effect of pulling a book to the floor after it has fallen off a table. It acts 24 hours per day, 7 days per week, 365 days per year. It is important and necessary, then, to ensure that these structural components have been properly designed and that they are adequate to resist the effects of gravity.

There are two primary considerations when considering gravity forces. The first can be thought of as "fixed" load and is commonly referred to as **dead load**. Dead loads are represented by the actual weights of the materials used in the building. For example, in a roof framing system (which in this example includes manufactured roof trusses; see Figure 1-6), the dead loads are the actual weights of the roof shingles, the felt paper, the plywood, the roof trusses, the insulation, and the **gypsum board** (wallboard) on the ceiling. These weights are always present and do not vary over time.

Likewise, the second consideration can be thought of as "variable" load. Variable loads, such as **live** and **environmental loads**, change over time and are not always present. For example, environmental loads such as snow and ice on a roof would be regarded as variable loads because they change with time. While they might be present in abundance in January, they will likely not be there at all in August.

Another type of variable load occurs with the floor and ceiling framing. With respect to floors, the live load from people and furnishings varies with time as the occupants of a dwelling move from room to room or as furniture is moved about. Inside an attic space, loads can vary as well. Think of all those holiday decorations in your attic: the artificial tree, the miles of lights, and even that fruitcake you planned to regift next year. These all make up what are known as live loads, because they may or may not be there and can change over time. The code establishes loading criteria for both fixed and variable loads to ensure that all structures are sound and capable of resisting gravity forces over time.

Unlike gravity loads, wind and seismic forces are not ever-present. That is, they occur periodically and often for short durations. Make no mistake though; they are both powerful forces that must be fully considered when a building is designed

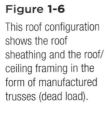

Figure 1-6

This roof configuration shows the roof sheathing and the roof/ceiling framing in the form of manufactured trusses (dead load).

structurally. Wind forces act horizontally on a building and have the effect of pushing the building from side to side. Additionally, wind forces can cause uplift in open structures or overturning if winds are severe enough. Seismic forces act horizontally or vertically, or both; and can cause significant damage to the dwelling and its contents.

Wind forces occur in every part of the country (and the world, for that matter) and can vary in intensity from the slightest breeze to hurricane-force winds. When the wind blows against a building, it pushes against the walls and roof, which in turn may cause the building to move from side to side. Under most conditions, this movement is imperceptible. Under high wind conditions, however, this movement is not only felt, but can also cause considerable damage.

These wind forces have the effect of bending or "flexing" structural members such as walls, roofs, and other components in a sideways movement. This loading occurs simultaneously with the gravity forces we spoke of previously and exerts tremendous stress on a building. As one might imagine, it is necessary to consider these loading conditions as well. The residential code ensures that the structural members used to build dwellings will resist all applicable loads and perform within safe limits to resist this bending.

Unlike wind forces, seismic forces do not exist in all parts of the country. Because they are not present everywhere, seismic forces are perhaps a bit more difficult to understand. They are certainly less intuitive than wind forces. Suffice it to say, if you live in an area of the country where seismic activity is not a part of your reality, it can be a terrifying event! For lack of a better description, imagine for a moment that you are standing on a small carpet. Then, imagine that someone grabs one end of the carpet and, in an instant, pulls it a foot or two away from you. It is easy to imagine yourself falling backward, or at the very least, teetering back and forth as you get your bearings. My first seismic event was very much like that.

Seismic forces are generated by earthquakes, which are caused when the earth moves beneath a structure. During these events, tremendous energy is released that must be absorbed by the structural components within the building. Earthquakes can cause a structure to move from side to side, like wind forces, and they can also cause it to move up and down (like riding a wave) or both. Special consideration of these movements must be given during the design of structures in areas prone to seismic activity.

One key consideration is to ensure that the house stays attached to the foundation while the structure endures the event. This takes special connections, as shown in Figure 1-7.

Figure 1-7

A tie-down, or hold-down, device used for resistance to seismic forces.

COURTESY DELMAR/CENGAGE LEARNING

These tie-downs or "hold-downs" keep framing members tied to the foundation and prevent uplift and overturning. Code-required anchor bolts prevent the structure from sliding off its foundation. In areas of high seismic activity, the residential code provides criteria to ensure that structures stay put during an earthquake.

Mechanical Systems

Mechanical systems, commonly called heating, ventilation, and air conditioning systems (HVAC), are provided in residential construction primarily to provide comfort for the occupants of buildings. These important systems provide protection from extremes of temperature and work in conjunction with the **building envelope** (insulated walls, roofs, and floors) to maintain a comfortable, tenable environment. HVAC systems not only deliver heated or cooled air to occupied parts of a dwelling, but they also provide fresh air ventilation and recirculate previously heated or cooled air for reuse (see Figure 1-8).

In very cold climates, for example, heating is typically the primary concern. In these types of climates, such as in mountainous regions and in the northern and eastern parts of the United States and throughout Canada, winters are often characterized by harsh, cold conditions. Summers, in contrast, are often mild and not usually subject to dangerous extremes of temperature. In these cases, HVAC systems add heat to the inside of residential structures, creating **conditioned spaces**. Conditioned spaces, as the IRC refers to them, are simply spaces in which a mechanical system is used to maintain an interior temperature of 68° F (20° C) measured at 3 feet (914 mm) above the floor as is required by Section R303.8 of the code.

Heating systems come in a wide variety of types and use a variety of energy sources. Forced-air heating systems blow air through **ducts** (rigid metal or flexible plastic "tubes") that deliver the conditioned air to a space or room within the dwelling (see Figure 1-9). Forced-air HVAC systems can be fueled by natural gas, propane, heating oil, or electricity. The IRC carefully regulates fuel gas piping to ensure that it is properly sized, properly supported, and is tested under pressure to ensure that it will not leak.

In any HVAC system where a fuel is burned, deadly carbon monoxide and other undesirable by-products are produced. These **products of combustion**, as they are referred to, must be properly vented outside of the dwelling to ensure that they do not accumulate to

Figure 1-8

A modern, gas-fired furnace and attached duct work.

COURTESY DELMAR/CENGAGE LEARNING

Figure 1-9

Ducts and a distribution box installed in an attic.

COURTESY DELMAR/CENGAGE LEARNING

dangerous levels inside the occupied space. The IRC also regulates fuel-burning appliances to ensure that the vents used to convey the products of combustion are properly sized and that they terminate properly outside the dwelling. Because these vents can often become quite hot during this process, the IRC regulates clearances to **combustible materials**, that is, materials that will ignite and burn when exposed to heat or flame, to ensure that proper clearances are maintained. This helps prevent fires from starting in the first place.

In addition to providing human comfort, another important function for HVAC systems is maintaining certain temperatures to ensure that other building systems function as intended. For example, when interior temperatures are properly maintained in cold weather climates, the risk that water pipes will freeze and burst in subfreezing temperatures is greatly reduced. Similarly, drain piping is less likely to freeze, thus keeping waste water and sewage draining properly from the structure.

Plumbing Systems

Plumbing systems are provided to ensure proper sanitation and facilitate storm drainage in a dwelling. They deliver **potable water** (water that is safe for drinking) to a dwelling through the use of plastic or metal pipes and also drain waste out of and away from a dwelling, also through the use of plastic or metal pipes (see Figure 1-10). Modern plumbing systems are essential to prevent the spread of disease and to maintain the health of both the occupants of the dwelling and the community. Plumbing systems also protect structures and property through the proper management of storm water. Plumbing systems are regulated under the plumbing provisions contained in Chapters 25 through 33 of the IRC.

Figure 1-10

Modern plastic plumbing piping shown in a water distribution system. This assembly is a manifold to which all water piping is attached for routing through the dwelling.

COURTESY DELMAR/CENGAGE LEARNING

Potable water is supplied from either a municipal source (such as from a city or water district) or an onsite source such as a well. Either way, water is delivered through the water service to the dwelling itself where it is used for drinking, cooking, bathing, and laundry. The water service consists of the water meter or well head, the service line (often called the water line) connecting from the meter or well head to the dwelling, and the water distribution piping contained within the structure. Typically, if provided through a municipal source, the water meter and all portions beyond it toward the water main are regulated by the water purveyor. All portions of the water service on the actual construction site (the dwelling side of the meter) are considered private and fall under the scope of the IRC.

One primary concern of the code is **cross-contamination**. This occurs when a potable water system is cross-connected to a nonpotable system and allows potentially contaminated water to mix with the potable water. Where this condition occurs, serious illness is possible if the water is contaminated with deadly bacteria or other toxins. For example, it is common for a landscape irrigation system to be tied to the water service piping. Without appropriate safeguards, the landscape irrigation system (an open system) can allow pesticides, fertilizers, pet waste, and other contaminants into the potable system.

Fortunately, the code regulates this condition and requires the use of an approved **backflow prevention device** (essentially, a one-way valve) to be used between the potable and irrigation systems. Such a device allows potable water to flow through the service piping and into the irrigation system, but will not allow water to flow in the other direction. These devices are highly effective at preventing cross-contamination.

Plumbing systems also drain waste and storm water from the dwelling. Like supply piping, waste and storm water piping—commonly referred to as drainage, waste, and vent (DWV) piping—is typically plastic but sometimes metal (see Figure 1-11). Some piping used for storm systems is also made of concrete. DWV piping drains sewage and **gray water** (waste from bathroom sinks and laundry trays) through the building drain and out away from the structure. Waste materials are either dumped into a municipal system called a sanitary sewer and transported through a series of pipes to a sewage treatment plant, or the waste is disposed of onsite in a septic tank and drain field. Onsite systems typically fall outside the scope of the IRC and may be regulated by other agencies through the adoption of the 2009 International Private Sewage Disposal Code or by local ordinance.

Figure 1-11

Drainage, waste, and vent (DWV) piping shown in a bathroom (double sink).

COURTESY DELMAR/CENGAGE LEARNING

These onsite systems are regulated heavily to ensure that they function properly and that they meet all applicable environmental laws and regulations.

Electrical Systems

Electrical systems are provided to ensure that electrical energy can be safely delivered to a dwelling. Electrical systems deliver electricity into a dwelling through the **service entrance** (point of entry) and use wires to distribute the electricity into **receptacles** (outlets), switches and lights, and other electrically operated equipment (see Figure 1-12). A code-compliant electrical system greatly reduces the risk of shock, electrocution, and fire. Electrical systems are regulated under the electrical provisions of Chapters 34 through 43 of the IRC.

Properly designed and installed electrical systems reduce the risk of fire in a number of ways. The **electrical service** (the main electrical panel) is the point of delivery into the structure and must be carefully sized to ensure that it can meet the demands of the modern dwelling without becoming overloaded. Overloaded systems generate significant amounts of heat that can ignite materials in and around the service, causing a fire. Circuits and breakers must also be properly sized to ensure that they do not become overloaded. The number of fixtures and receptacles on each branch circuit are carefully regulated to ensure that this does not happen. Wire size is carefully regulated as well. Generally, the greater the load to the circuit the larger the wire diameter must be, as the larger wire size is capable of delivering more current than smaller sizes. The larger wire dissipates heat better than the small wire as well.

One primary concern of the code is **grounding**. Grounding is a critical safety feature in an electrical system because it provides a "path of least resistance" when a short occurs. A properly designed, correctly installed ground is one of the most fundamental and important life safety features in a dwelling and, therefore, is heavily regulated by the code. This important life safety feature aids in the prevention of electrocution because it allows electricity to flow to the "ground" rather than through the human body (see Figure 1-13).

The code further protects against electric shock by requiring all receptacles located outdoors and in garages, unfinished basements, kitchens, and similar areas to be protected with ground-fault circuit interrupters (GFCI). These special receptacles interrupt the flow of electrical current when they sense a ground fault from

Figure 1-12

An electrical service panel and adjacent meter base for a dwelling unit.

a defective appliance or other equipment. Recent code changes also require arc-fault circuit interrupters (AFCI) to be provided in dwellings. AFCI shut off a circuit when the building wiring or an electrical cord plugged into a protected circuit is damaged, causing arcing or sparking due to intermittent current flow. This newer technology aids in the prevention of electrical fires.

THE ROLE OF THE BUILDING OFFICIAL

The building official (also known as the code official) is responsible for the administration of the residential building code. Throughout this text, these terms will be used interchangeably. Oregon law requires a building official to be named if a county or municipality is to administer its own building code program. In other areas, the requirement for a building official is created when the IRC is adopted, usually by local ordinance. Upon adoption, Section R103 of the code creates a department of building safety and directs that the official in charge of such department is known as the **building official**. Either way, the building official has an important legal responsibility in the enforcement of the residential code.

Essentially, the building official is responsible to the community in which he is authorized to administer the code to ensure that the built environment is safe and code compliant. This is an enormous responsibility and not one to be taken lightly. Day in and day out, the building official must be concerned with the built environment in ways we are often unaware exist. In fact, most people do not even consciously think about building safety.

Why? Because those of us who are fortunate enough to live in areas where recognized building codes are adopted *and* vigorously enforced enjoy unparalleled building safety. Don't think so? Take this simple test: The last time you visited your local, modern mega-mart or department store, did you stop and wonder upon entering whether the roof would collapse or if there were adequate exits? Probably not, because the building official does his job to ensure that building codes are properly enforced. Through rigorous enforcement, he and his deputies ensure that buildings are properly designed and built so they will not collapse under the weight of the accumulated dead and live loads and that exits are sufficient in quantity and properly spaced so they can be used quickly and safely in an emergency. While it is easy for us to take building safety for granted, the building official cannot do so by his charge.

Figure 1-13

A rebar stub onto which the electrical grounding system is attached.

COURTESY DELMAR/CENGAGE LEARNING

Duties and Responsibilities

Section R104.1 of the IRC not only authorizes the building official to enforce the provisions of the code, but also *directs* him to do so. Thus, code officials have a serious and vital role in the community, and an important job to do. They must ensure that all aspects of the code have been considered when reviewing the construction plans and when the project is constructed.

This same section also grants the building official authority to render interpretations of the code, which means that he can decide what is meant by each provision of the code and how it is to be applied. Although the code is certainly comprehensive, it cannot possibly address all possible circumstances that might occur during a construction project. Thus, he must be given the authority to use his judgment and to consider all the facts in each case. This allows him to deal effectively with the "gray areas," that is, those areas where the provisions of the code do not quite fit the specific conditions or circumstances presented in the project.

The building official can also adopt policies and procedures that enable him and his staff to effectively administer the code. This means that he has authority to create policy direction for his department so that he and his staff can provide consistent service and enforcement. It is important to note, however, that such policies and procedures must be "consistent with the intent and purpose of the code." Also, the policies and procedures established by the building official cannot waive the specific requirements of the code. For example, since the code specifically requires smoke alarms in bedrooms, it would be a violation of Section R104.1 to create a departmental policy that waives such requirement. It is wise and highly recommended to know and understand the policies and procedures established in the jurisdiction where the green building project will be built. Doing so will assist greatly in the quest to go green!

The Building Official and You

The building official in each community is, by nature, a cautious individual. He has to be, given the enormous responsibility he carries in the administration of the IRC. Thus, the relationship with him, his deputies, and others associated with the local building department must be founded on trust, confidence, and good faith. And, as with most other human interactions, strong relationships are fundamental to the success of any project. This is true in the business world, in our personal relationships, and of course in the world of building codes. The following are effective ways to develop and strengthen the relationship with the building official in your jurisdiction.

Communication

Ben Jonson, English playwright and poet, once said, "Language most [shows] a man. Speak, that I may see thee." I have always loved this quote because I think it captures the essence of communication. What we say and how we say it speaks volumes about our character, what we know and, essentially, who we are. This quote is shared with you not so much as a lesson in communications but to reinforce the idea that communication is an essential part of any human interaction.

This is especially true in the relationship with the building official. Communication is a fundamental part of the permitting and construction process so it makes sense to really grasp and understand why it is important and how it can help move toward a successful project. In my career as a code official, there have been many instances of successful communication. There have also been, unfortunately, many instances of absolutely dismal communication, some of which caused great expense and heartache for those involved. From these experiences, the following suggestions are offered.

Avoid surprises: No one likes to be taken by surprise. This is especially true for building officials. When communicating with him, tell the entire story. Do not leave something out because it seems trivial or may make matters worse. When something goes wrong, which invariably it will, be upfront about it. There have been numerous occasions in my career where code-related decisions have been made based on information *believed* to be complete and accurate, only to discover later that one key piece of information was left out—one that changes everything! The building official cannot help you if he does not know and understand the whole story.

Speak the language: As in any industry, the more effectively one speaks the language of the code official, the more successful one's communications will be. The more time spent knowing and understanding the terminology used in the code, the easier it will be to understand each other. One of the main reasons I wrote Part I of this text is to prepare homeowners, builders, and design professionals to speak the language of the code official and those with whom they will be dealing during the green construction project. Invest the time it takes to learn to do this well and take it seriously. It will pay big dividends!

Tell the truth: Enough said.

Confidence

Fundamentally, a green building project is much more likely to be successful (or any building project for that matter) if the building official is confident in one's ability to meet the requirements of the code. The better one understands the duties and responsibilities of the code official, the departmental policies he has established for effective code administration, and the code itself, the more confident he can be that the homeowner, builder, or design professional will deliver a code-compliant building.

A Shattered Dream...

To further illustrate the importance of communication with the code official and the potential consequences of failures in this important area, I offer the following real experience.* In early 2001, I was contacted by a couple who were devastated to discover that their dream of converting their beautiful, newly built home and multiacre property into a home-based business—a wedding facility—had been shattered.

The Smiths had several years prior begun construction for their new dwelling. Their vision was to construct a lavish dwelling on their multiacre property in which they would not only live but would operate a home-based business in the growing wedding service industry. Their dream was not only to eventually retire in this lovely setting but also to share their beautiful home and picturesque, wooded property and gorgeous views with those who sought their services. The plans for the dwelling included several large rooms on multiple levels that were sized to accommodate large wedding services as well as food service and reception space for as many as 100 guests.

A few months after the dwelling was completed and the Smiths were settled into their new home, they came in to apply for the necessary permits and to obtain land use (zoning) approval, the final step in the fulfillment of their dream to opening a wedding facility, or so they thought. They filed the necessary applications with Planning and Zoning and some weeks later received the necessary land use approvals to operate the wedding facility as they had planned. Upon receiving the approval, they were also referred to my building department to ensure that the structure they intended to use in the operation of their business met all applicable building codes. It was upon this first meeting with me that their dreams began to unravel.

It seemed that, although the Smiths had gone to great lengths to inform their contractor of their ultimate plans, he failed to share this critical information with my building department at the time of application. When the contractor had the plans prepared and filed the permit application, he indicated only that the scope for the project included the construction of a single-family dwelling. The builder assumed that the owners' plans to use portions of the new dwelling and their residential property as a wedding facility could be easily dealt with later and never mentioned this fact at the time of permit submittal.

As a result, my building department performed a plan review, approved the plans, issued a building permit, and performed all the required inspections with the belief that the project would be used as an ordinary dwelling much like any other residential construction project. When the building passed all final inspections, we allowed occupancy based on the approved use as a dwelling, not as a mixed occupancy residential/commercial use building as their plans required.

Sadly, upon our discovery of the Smiths' plans to use portions of the building for wedding services and receptions, we realized their dwelling could never be used in its present form as they had intended. Under the code provisions in effect at that time, the areas within the dwelling intended for commercial use should have been separated from the residential portions by fire-resistive construction. The types of materials used to construct the building were not compatible with the proposed commercial use.

Additionally, local regulations required an automatic fire sprinkler system to be provided throughout the structure which, of course, had not been installed. Had we known upfront of the plans to use portions of the dwelling for wedding services and receptions, we could have advised the owners of these code requirements and the steps they could have easily taken during construction to

(continued)

A Shattered Dream... (continued)

ensure that their dreams could be fulfilled. Repairs and alterations necessary to convert the structure after the fact would have cost more than $100,000—money the Smiths did not have.

Ultimately, the Smiths sued the builder and spent nearly 7 years embroiled in a bitter, costly legal battle. They did eventually win damages; however, they had to make significant compromises in the operation of their business because the obstacles related to the construction could not be overcome. More important, the Smiths never fully realized their dream because critically important information was left out during the review and permitting processes.

Although this is certainly an extreme example of the possible consequences of a miscommunication, it is a very real story. It proved to be not only very costly for the owners of the property but heartbreaking as well; and the story has played out in many variations over the course of my career. Make sure the code official knows and fully understands your full intentions and the full scope of your project so that he is in the best possible position to be of service to you. Yes, his message may be difficult to hear initially, but it is far better to know the full requirements upfront and make informed choices at the design stage than to be surprised later—or worse, to find out that your dreams are impossible to achieve.

*Although based on a true story, the names of the owners and the specific circumstances surrounding the case have been fictionalized for the purpose of this book.

Also, the more confident the building official can be in the homeowner's and builder's knowledge of the code and their grasp of its various provisions, the more likely he is to work cooperatively and approve the project. As mentioned previously, this confidence will pay big dividends in the development of the green building project.

THE ROLE OF THE PLANS EXAMINER

The **plans examiner** (sometimes known as a plan reviewer or plan check engineer) has an important role in the permitting process. Section R103.3 of the IRC allows the building official to appoint deputies to assist in the administration of the building code. Thus deputized, the plans examiner has a supporting role in the enforcement of the codes. Where the building official provides policy and administrative guidance in the administration of the code, the plans examiner provides a practical, hands-on role prior to construction in that he reviews the project before it is built. This provides an opportunity to catch major errors *before* a structure is built, to prevent time-consuming and expensive repairs.

As the name implies, the plans examiner *examines* or reviews construction drawings to ensure that the work shown on the plans, if built as proposed, will be code compliant. It is his job to review each of the various parts of the building plans (e.g., floor plans, foundation plans, framing plans) to determine if the pictorial representation of the proposed dwelling depicts construction that will meet

Figure 1-14

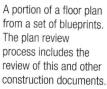

A portion of a floor plan from a set of blueprints. The plan review process includes the review of this and other construction documents.

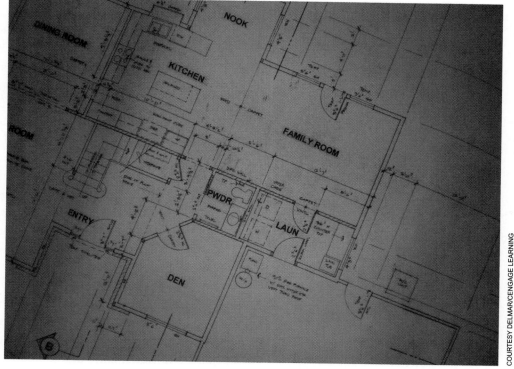

the requirements of the code (see Figure 1-14). He will often perform a structural review to determine that all structural components and systems are sound and adequate to resist the required design loads imposed by the code.

He will also perform a life safety review to determine if the plans correctly show required items such as emergency escape and rescue opening, smoke alarms, and code-compliant stairs and guards. Additionally, he will check the plans to determine that they are clearly detailed and provide the necessary information to allow the dwelling to be built and inspected. The notes and specifications must clearly show lumber grades and other related information so that those who construct the building know what materials to use. He will also perform other reviews, such as for energy efficiency and, in some areas, to determine compliance with the mechanical, plumbing, and electrical portions of the code.

THE ROLE OF THE INSPECTOR

The **inspector** (sometimes called a building, mechanical, electrical, or plumbing inspector) also has a supporting role. Like the plans examiner, he is deputized by the building official to perform those duties assigned per Section R103.3. Also like the plan examiner, it is his job to support and enforce the policies of the code official, an equally important job. One key difference, however, is that the inspector's work is performed in the field while the dwelling is actually being built. By performing periodic inspections, the inspector ensures that the structure is built in accordance with the approved plans and that it meets all applicable codes.

An inspector performs visual inspections during the construction process. It is his job to inspect footings and foundations, framing, electrical wiring, plumbing systems, insulation, and other items to ensure that all work installed complies

Figure 1-15

The interrelationship of the code administration, plan review, and inspection processes. Effective code administration requires both plan review and field inspection to ensure a code-compliant structure. Plan review ensures that the proposed construction will meet current codes prior to being built. Field inspection ensures that the construction is built according to the approved plans.

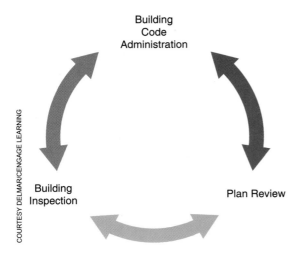

COURTESY DELMAR/CENGAGE LEARNING

with the requirements of the code. In some areas, these inspections are performed by the same person and in other areas they are performed by specialists with training in each major area of construction. Inspectors provide field reports that identify deficiencies in the construction as it progresses and follow up to ensure that the corrections are made. At the end of the project, inspectors perform what are called "final" inspections. These inspections are the last step in the construction process. During the final inspections, inspectors make sure all prior corrective work has been completed and determine whether the dwelling is safe to occupy. If this is the case, a Certificate of Occupancy is issued which states that all aspects of the code have been met and that the structure can be occupied.

Collectively, the building official, plans examiner, and inspector make up a three-part team that leads to effective code enforcement and safe communities. Each of the three is dependent upon the others and if any one part is missing, code enforcement on the whole suffers (see Figure 1-15). For there to be effective code administration, plans must be reviewed prior to the start of construction to avoid expensive corrections and delays on the field. Additionally, construction must be inspected periodically to ensure that the structure is actually being built according to the approved plans and that **field changes** (changes due to unforeseen circumstances or someone simply changing one's mind) are also code compliant. Without plan review and/or inspection, a large hole exists and there can be no effective administration.

LOOKING AHEAD

Whether you are a homeowner, builder, or architect, you are beginning to have an understanding of the importance of the residential building code and the people who enforce it. You should also have an understanding of the duties and responsibilities of those who administer the code and the important jobs they do. Let us now explore the code itself so that we can more fully understand the way construction materials and methods are regulated. This understanding will provide the necessary background to frame green building materials and construction methods in the context of the code and move us closer to going green with the IRC.

CONSTRUCTION MATERIALS, METHODS, AND THE IRC

2

learning objective

To know and understand how construction materials and methods are generally regulated in the International Residential Code so that this knowledge can be later applied to green building materials and methods.

BUILDING CODE FUNDAMENTALS

In the first chapter, we learned about why we have building codes and how they work to keep us safe in our homes. We also learned about the people who work with the building codes and the enormous responsibilities they have. We will now explore specifically *how* the IRC regulates construction, as this will move us forward in our quest toward going green with the IRC.

There are two principle ways in which the IRC regulates construction: construction materials and construction methods. In the case of construction materials, the IRC sets standards for the basic materials to be used. These materials are commonly (but not limited to) wood, concrete, masonry, insulation, and gypsum board. For our purposes, we will also include mechanical equipment (furnaces, heat pumps, exhaust fans, etc.) as well as regulated appliances (cooking stoves, ovens, fireplaces, etc.) in the category of construction materials as these are regulated as well.

In the case of construction methods, the IRC prescribes how it is we are to put the basic materials together. That is, it provides us with instructions that, if followed, will result in a code-compliant structure. Whether we use wood, concrete, masonry, or various combinations of these and other materials to build a dwelling, the IRC provides guidance so that we can assemble the materials in a safe, code-compliant way. The IRC also provides us with "instructions" in other areas as well, such as in the categories of life safety and energy efficiency.

Before we explore these topics, however, there are a few basic code principles that we must understand in order to more fully appreciate the language of the IRC. These principles are true across all chapters in the code, regardless of which topic or code provision you are reading.

- The IRC establishes both general and specific code requirements. Where conflicts occur between the two, the *specific* code provision always takes precedence over the general. The IRC may also specify different materials or methods of construction in different code sections. When this occurs, the most restrictive requirement governs.
- The IRC uses the terms "shall" and "may" throughout the code. The word "shall" indicates a mandatory code requirement. In these cases, the permit holder has no option but to comply with the provision unless an approved alternative method or modification has been approved by the code official. The word "may" is permissive language. In these cases, the IRC usually offers a number of code-acceptable options from which the builder or homeowner can choose.
- Footnotes often appear below tables and figures that offer critical information. These footnotes apply only to the table or figure to which they are associated. More important, they can significantly alter the code requirements outlined in the table. Read all footnotes carefully, because in these cases, it really is the fine print that can get you.

CONSTRUCTION MATERIALS AND THE IRC

Fundamental to the IRC is the way in which it handles construction materials. We learned previously that materials used in or on buildings must be durable; capable of withstanding loads imposed by gravity, wind, or seismic forces; and in some cases fire resistant. Additionally, building materials must perform as intended over the reasonable life of the structure. This is true of traditional materials as well as green building materials.

Building materials fall into a variety of categories. Some materials are derived directly from sources in nature. Lumber, for example, is milled from logs that are cut from particular species of trees felled for use in construction. Other building products are assembled from combinations of naturally occurring materials and man-made materials, such as concrete. Concrete is made from varying proportions of water, **aggregate** (course gravel and sand), and cement (a man-made material). Still others are manufactured entirely from man-made materials such as the plastic pipe used in plumbing systems. Whether derived from a naturally occurring source or manufactured, all building products must meet the requirements of the IRC.

How does the code accomplish this? It does so by establishing minimum performance criteria based on widely accepted standards. Internationally recognized organizations such as the ASTM International (formerly known as the American Society for Testing and Materials), the National Fire Protection Association (NFPA), and Underwriters Laboratories (UL) provide criteria and standards of performance that ensure building materials, equipment, and regulated appliances will comply with specific code provisions and meet the intent of the code. The IRC specifies very clearly how each product, system, or appliance must be determined to be suitable for use under the terms of the code. Standards can dictate how materials and building systems are to be tested, how they are to be manufactured, and how they are to be designed or installed.

Referenced Standards

Section R102.4 of the IRC, Referenced Codes and Standards, establishes that all standards referenced in the code shall be considered to be a part of the code to the extent that particular standard is prescribed. It might be the case that an entire standard is referenced, or only a portion of a particular standard is referenced. Either way, it is the combination of the code text itself plus the referenced standards that forms the complete IRC.

Standards are published by a variety of organizations. Some of these are professional associations or societies with an interest in the promotion of a particular field. Others are public-interest-based institutions whose efforts focus on public safety, health, and welfare. Either way, organizations that develop and publish standards are highly respected and sought after for their expertise and varied interests in a particular industry.

Many of these standards are developed using an open consensus process to ensure that the standard is developed fairly. Some organizations assemble committees made up of a broad cross section of society consisting of industry representatives, public members, and regulators in an effort to hear from all groups with an interest in the subject at hand. These efforts—though time consuming and sometimes challenging—contribute to the publication of standards that are fair and in the public's best interest.

Some of these groups or their affiliates conduct extensive research and often perform testing on building products and systems under a variety of conditions. Although they are often simulated, these test conditions are designed to mimic those that would typically be encountered over the life of a structure. Standards-writing organizations employ or work closely with engineers, research scientists, and academics with extensive training, education, and experience, some of whom are internationally recognized as experts in their field. Many of these organizations have been conducting or compiling research for decades and have amassed an astounding amount of knowledge.

Standards appear in a variety of formats including books, technical publications, reports, and in other printed materials. Publications produced by these organizations often contain the results of tests performed on various building materials and products as well as analysis and recommendations for the use of the material in question. These publications provide the scope, context, and specific conditions under which a particular product should be used as well. More important, they provide limitations for the use of a product and identify conditions under which the use of a product or system would be prohibited. In short, they provide comprehensive, expert guidance into the use of a particular material or system. The following illustrate how standards are applied to a variety of traditional building materials.

Figure 2-1

Sawn lumber used
to frame walls and a
stairway in a dwelling.

Sawn Lumber

A traditional building material for residential construction, lumber is used extensively throughout the United States and Canada. Lumber is sawn from wood logs (hence the name) into framing components of varying sizes for use as wood studs, floor joists, ceiling joists, roof rafters, posts, beams, and timbers (see Figure 2-1).

The ability of a particular piece of lumber to resist a load varies based on a number of factors. Some of these factors are external and depend upon the particular design conditions in the construction. For example, the size of the structure, the number of stories supported, and the roof framing configuration all affect how far the lumber will span and how much weight it will carry. Geologic and geographic conditions also affect the structural design and must be carefully considered for obvious reasons.

Some factors, however, are inherent to the particular piece of lumber itself. The species of the tree from which it was sawn, the grade of the material, and the size of the framing member all greatly affect the strength of the wood and its ability to carry a load. These inherent or "internal" properties are known as **structural properties**; and because the load-resisting capabilities of each piece of lumber and other wood products are directly related to these structural properties, it is crucial that the species and grade of each piece of lumber be properly identified. This is typically accomplished through a **grade stamp** (an indelible ink stamp) applied to the surface of a piece of lumber or other wood product such as plywood, oriented strand board (OSB), or particleboard sheathing (see Figure 2-2).

Figure 2-2

A grade stamp and identification markings on floor sheathing. Note the sheathing thickness (23/32), span rating (24 O.C.), adhesive type (exterior), and other information provided in the stamp.

COURTESY DELMAR/CENGAGE LEARNING

Wood studs are typically installed vertically and are used to form exterior and interior walls. They vary in size but are usually found in 2 inch by 4 inch (51 mm by 102 mm) and 2 inch by 6 inch (51 mm by 152 mm) nominal sizes. Wood studs are graded visually and under the 2009 edition of the IRC, studs are generally required to be a minimum of No. 3, standard, or stud grade lumber per Section R602.2 (see Figure 2-3).

Wood floor joists, beams, and girders are typically installed horizontally and are used to form floors. Like wood studs, they vary in size but are usually found in 2 inch by 10 inch (51 mm by 254 mm) and 2 inch by 12 inch (51 mm by 305 mm) nominal sizes. Although there is no minimum grade established for floor joists per se, the 2009 IRC generally requires floor construction to be capable of resisting all loads and also to transfer those loads to supporting structural elements. Additionally, Section R502.2 requires all floor construction to conform to the requirements of Chapter 5 of the IRC or, alternately, with the requirements of the American Forest and Paper Association (AF & PA) National Design Standard, a widely used standard in the industry.

Similarly, wood rafters, trusses, and ceiling joists are installed in either sloped configurations (as with a roof) or horizontally, as with a flat ceiling. Again, sizes vary but are commonly found in 2 inch by 6 inch (51 mm by 152 mm), 2 inch by 8 inch (51 mm by 203 mm), and 2 inch by 10 inch (51 mm by 254 mm) nominal sizes. Although there is no minimum grade established for rafters and ceiling joists, the 2009 IRC generally requires roof and ceiling construction to be capable of resisting all loads and also to transfer those loads to supporting structural elements just as it does with floors. Additionally, Section R802.2 requires all roof and ceiling construction to conform to the requirements of Chapter 8 of the IRC or, alternately, with the requirements of the AF & PA National Design Standard.

Concrete

Concrete is used in modern residential construction in a variety of ways. Most frequently, it is used to form the footings and foundations for dwellings, as well as to form slabs for garages and basements (see Figure 2-4). Sometimes, concrete is used to form the walls of the dwelling itself and can occur above grade or below, such as with a basement. Concrete is an excellent material for use in foundations because

Figure 2-3

A grade stamp on a wood stud. Note the species (D Fir) and grade (No. 2).

it is incredibly hard, a property known as **compressive strength**.

Concrete is typically of the poured-in-place type in residential construction. That is, plywood forms or "molds" are placed on the jobsite. Concrete is mixed in batches and delivered to the site where it is poured into the forms. It later hardens and provides a suitable base on which to build the rest of the structure.

The ability of concrete to resist a load is directly related to the way in which the raw materials are mixed and other factors. Certain proportions of cement (commonly Portland cement), aggregates, and water are mixed together, and it is these proportions that determine the concrete's strength. Additionally, other products are sometimes added (called **admixtures**) that further improve the concrete's strength or change the way in which it cures. Although concrete has excellent compressive strength as previously mentioned, it is relatively weak in terms of **tensile strength**. That is, its structural properties are relatively low when concrete is loaded in tension. Tensile forces occur when concrete is stretched rather than compressed, such as occurs in the bottom of a footing or concrete beam as it flexes to resist the load. As a result, reinforcing steel is typically added, creating a composite material that can resist both compressive and tensile forces.

Section R402.2 of the IRC establishes minimum compressive strengths for concrete. In most cases, the minimum compressive strength is 2,500 pounds per square inch (psi) (17,237 kPa); however, in areas prone to moderate or severe weathering, the minimum strength is 3,000 psi (20,684 kPa) to 3,500 psi (24,132 kPa) depending on type of construction. The code requires the materials used in concrete and the test methods used to determine its compressive strength to meet the requirements of the American Concrete Institute Standard 318, Building Code Requirements for Structural Concrete (ACI 318).

As we discussed earlier, the ACI is a perfect example of the type of organization that develops the standards referenced in the code. The ACI has been in existence since 1904, and is dedicated to the advancement of knowledge of concrete structures. With more than 100 years of experience in the study of concrete, ACI has unparalleled expertise in the industry. Its publications are an industry standard and it can be easily seen why these standards are referenced in the IRC.

Figure 2-4

Concrete used to form stepped foundation walls on a sloped lot.

BUILDING PRODUCT EVALUATION REPORTS

For the code official, the most complete and reliable method of demonstrating the adequacy of a building product is known as a product evaluation report. Also known generically as a "product approval report," these reports are produced by the International Code Council Evaluation Service (ICC-ES) as an assurance to the code official that a particular construction material or building system is code compliant and can be safely used in residential construction.

The ICC-ES is a nonprofit, public benefit corporation that provides technical evaluations of building products, components, methods, and materials. Product evaluation reports are produced at the request of a product manufacturer that seeks to have a specific building product or system evaluated. The ICC-ES works jointly with a manufacturer to evaluate a proposed material, component, or system against established criteria and, ultimately, the building code, resulting in the issuance of a report that provides clear evidence that the product or system meets the code.

More than simply providing the results of test data, evaluation reports provide a complete picture for the code official regarding the use of a particular material in construction (see Figure 2-5). These reports certainly address the results of tests performed on the products, but they also provide specific conditions of approval for the use of the products and establish clear limitations to ensure that the products are used as approved. These evaluation reports also provide strict identification requirements to ensure that the approved products are clearly labeled and can be easily identified on the jobsite. In the following sections, let us more fully explore the product approval process.

Understanding the Building Product Evaluation Process

As stated in the previous section, the issuance of a building product evaluation report does not mean that a manufacturer simply provided the results of any old test for review. In order to obtain an approval report, the building product in question must pass the *right* tests in a carefully documented process and with considerable oversight. Why? Well, without regulatory oversight or rigorous, uniform standards for quality, it is possible that a manufacturer or fabricator could find a

Figure 2-5

ICC-ES evaluation reports (see Appendix B for more information).

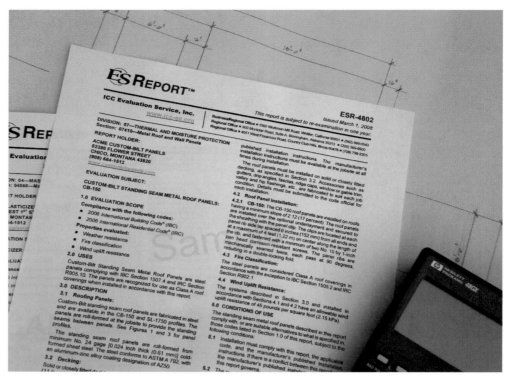

lab or agency that would provide "just the right results" to ensure that their products looked acceptable to code officials and consumers.

In fact, without oversight, manufacturers could perform their own testing, the results of which would always be at question due to the organization's financial interests in the results. Because this financial interest creates such an inherent and obvious conflict of interest, it is necessary that the process used to evaluate construction materials and systems be fair, unbiased, and credible. The product approval process is, then, much more complex than simply hiring a laboratory to perform various tests.

For a product approval to be meaningful, great faith must be placed in its findings and recommendations for use. Consider the needs of those who rely on them: The homeowner must be assured that construction products will be reasonably durable and perform the way they are represented by the manufacturer, providing a good value for the money invested in the dwelling. Builders must be certain that the materials they use will be of reasonable quality, be warrantable, and minimize call-backs once installed. Architects and engineers must be confident that the products they specify will perform as intended, meet project specifications, and fulfill their professional duties. Code officials must be assured that the products used in the construction of the dwelling are safe, structurally sound, fire resistant, and fully code compliant.

It is understandable, then, that the product approval process must establish a high bar to ensure that the findings presented in evaluation reports are reliable and meaningful, and that ultimately the products and systems approved for construction will meet all applicable standards for quality and performance. The next few sections establish how this is accomplished.

Proceed ... with Caution!

As the following true experience will demonstrate, I am certainly no exception when it comes to the "code officials are skeptics" rule. Allow me to share this example from my early career to further illustrate how building product evaluation reports can calm the fears of even the most cautious of code officials.

In the early 1990s, I began to encounter what was then a relatively unknown building framing product during the course of my daily inspections. The framing material was used primarily in floors but could also be used in roof framing applications. The product was lightweight, strong, and could span relatively long distances compared to sawn lumber. The framing material was also innovative, as it appeared to be manufactured from what would otherwise be considered to be waste or scrap wood products. What was the product, you ask? Wood I-joists, of course!

The manufacturers of these products touted their many benefits including longer joist spans; straighter, truer joists; quiet floors; and what was, in their opinion, overall superior performance when compared to sawn lumber. Because the joists could span longer distances, this also meant fewer intermediate supports (i.e., posts and beams) were required, allowing for larger open spaces. At that time, however, I was convinced that nothing could beat traditional sawn lumber. After all, hadn't traditional sawn lumber been used for decades with great success? Why change a good thing, right?

Not having encountered such a material before, I had many questions regarding its use in construction. How were the allowable spans determined under given loading conditions? How should the material be supported and braced? What types of fasteners should be used? In what framing application should this product be used? Where should it *not* be used? At that time, I was very concerned that the product would not meet the requirements of the code and I was also skeptical about the product's ability to perform to the levels suggested by the manufacturers. As a result, I routinely asked for engineering from the manufacturers to demonstrate that their products would work as intended.

As requested, the wood I-joist manufacturers willingly provided the necessary engineering for review. Eventually, however, these manufacturers also provided evaluation reports, detailing the extensive testing performed on their products and outlining the conditions under which their products could be used. The evaluation reports described the use of the product in detail and also answered the many questions I and other code officials had at that time. These reports provided span tables similar to those found in the code for conventional sawn lumber. They also identified conditions under which the I-joists could and could not be used, and specified fastener types, bracing requirements, and the like. In short, the reports answered all my questions and provided the necessary assurance that the products performed as stated.

The existence of evaluation reports marked a significant shift in my thinking regarding wood I-joists. In fact, the shift was so great that I eventually used wood I-joists in the construction of my own house! Once the evaluation reports were provided, I and other code officials were immediately assured that the products not only met all applicable code requirements but were given the guidance needed to review and inspect these materials with confidence. The concerns I had then seem almost comical today, because wood I-joists have gained so much popularity over the years that it would be difficult to find a dwelling under construction that does *not* include them.

Acceptance Criteria

As we saw earlier, the residential building code often provides clear criteria for evaluating the quality and suitability of certain building materials. For example, the evaluation criteria for sawn lumber used in walls is clearly and sufficiently spelled out in Chapter 6 of the IRC. Here, species, grade, size, and so on are spelled out in the actual provisions of the code.

There are times, however, where the code provisions and referenced standards are not sufficiently clear to address a particular building product or system, or where these products and systems fall entirely outside the provisions of the code. In these cases, the building materials are regarded as alternates to the code and criteria must be established, where possible, so that they can be considered suitable under the code.

To ensure a fair process for evaluation, the ICC-ES establishes acceptance criteria as the basis for the evaluation of a building product or system. This establishes a level playing field—a standard if you will—so that all manufacturers who make a similar product are evaluated against the same criteria.

Acceptance criteria (AC) are developed by ICC-ES technical staff in consultation with the manufacturer seeking product approval. These criteria are developed through the open, public hearings of the ICC-ES Evaluation Committee. This committee is made up entirely of code officials. Industry and other interested persons participate in the hearings process and provide input for consideration by the committee. Once approved, the acceptance criteria are published on the ICC-ES website and are available for all interested persons to view at no cost.

Testing to Recognized Standards

A significant part of most acceptance criteria is the establishment of standards to which building materials and systems must be tested. These tests are typically industry standards and are based on widely known and accepted test methodologies identified by, most notably, organizations such as ASTM International, the NFPA, and Underwriters Laboratories, but also organizations such as the American National Standards Institute (ANSI) and the Engineered Wood Association (formerly the American Plywood Association or APA).

Such criteria provide the manufacturers and fabricators of building products with a "target" to hit. That is, they clearly establish the specific, standardized tests that their products must pass if they are ultimately to be found acceptable under the code. In so doing, the manufacturer can set about the task of having its various products tested for strength, durability, and weathering, among other qualities. These test results help them to refine their products until they satisfactorily meet the requirements of all identified tests in the acceptance criteria.

One such test, for example, is the ASTM E84 Standard Test Method for Surface Burning Characteristics of Building Materials. This venerable test has become the classic method for testing the rate of flame spread and the density of smoke generated when a flame is applied to building materials, and is widely known in code enforcement circles (see Figure 2-6). It is specified as a required test in a great many of the acceptance criteria published by ICC-ES. Other commonly specified tests are the ASTM E119 Standard Test Methods for Fire Tests of Building Construction and Materials and the NFPA 286 Standard Methods of Fire Tests for Evaluating Contribution of Wall and Ceiling Interior Finish to Room Fire Growth.

Figure 2-6

The Steiner Tunnel Furnace used in the ASTM E84 tunnel test.

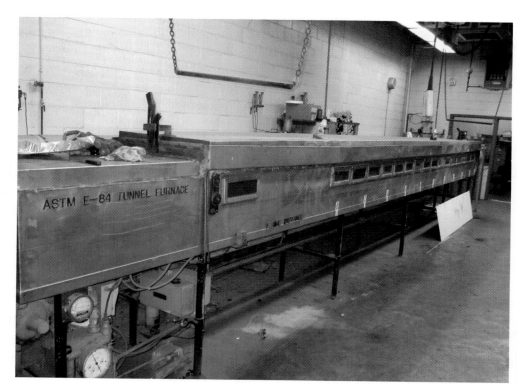

COURTESY OF HPVA

Approved Laboratories

Laboratories that perform product testing on building materials must generally be accredited by a recognized agency. This typically means that they must either be accredited by the International Accreditation Service (IAS) or by another accrediting body that is a signatory to the Mutual Recognition Arrangement (MRA) of the International Laboratory Accreditation Cooperation (ILAC). Like the ICC-ES, the IAS is a subsidiary of the ICC and is an independent nonprofit, public benefit corporation that accredits testing and calibration laboratories, inspection agencies, fabricators, and building departments, providing an assurance that these organizations deliver reliable, quality services.

Additionally, each lab must be accredited for the specific test method in each specific area it wishes to perform testing. For example, if a lab wishes to perform testing on the compressive strength of concrete, it must be found to be competent in this area and specifically accredited for that type of testing. This stringent requirement ensures that a lab has the necessary skills, training, education, and equipment to perform the tests against the evaluation criteria required for approval.

Third-Party Inspections

Before an evaluation report is published, it is necessary for an inspection of the manufacturing facility to be performed. Termed *manufacturer qualifying inspections,* these inspections are an important and necessary part of the approval process because they provide independent verification that the product being manufactured is consistent with that specified in the report and that the manufacturer has the means, equipment, and manpower to produce the product in accordance with the report. But the intent of the inspection is twofold, as it also ensures that the manufacturer has documented and implemented a quality system meeting the acceptance criteria outlined in AC10 (see the next section).

Because the qualifying inspection is so important and because it carries with it such a high level of responsibility, the ICC-ES provides detailed criteria for third-party inspection agencies to perform its duties. These criteria establish clear expectations and ensure fair, consistent inspection practices for both the initial and ongoing inspections. The duties and responsibilities of the inspection agency are provided in the ICC-ES Acceptance Criteria for Inspection Agencies (AC304).

Like testing laboratories, agencies that perform qualifying inspections must also be accredited to perform inspections for the particular products that are the focus of the evaluation report. It stands to reason that such an agency should have the necessary skills, training, and experience to evaluate the product fairly and objectively. Additionally, the inspection agency must also be accredited to those criteria found in AC304.

Quality Control

Like third-party inspections, quality control (QC) is also an important part of the product approval process. An approved quality control program is essential if a manufacturer seeks approval of its building products. A well-prepared, thorough plan is the way in which a building products manufacturer ensures that the materials being produced are of consistent quality and that they are manufactured in the same way as those materials used in the testing of the products.

Just as important as the plan itself, quality control must be carefully documented as well, by both the manufacturer and the inspection agency performing the third-party inspection. This documentation is ultimately what will be reviewed by ICC-ES staff as they evaluate a product for approval. ICC-ES Acceptance Criteria for Quality Documentation (AC10) provides the necessary guidance for anyone who manufactures a building product and seeks approval. The manufacturer seeking approval uses Appendices A and B of AC10 to gather and summarize data relevant to the ICC-ES approval. The inspection agency completes Form Q-21 during the factory inspection and the ICC-ES staff utilizes Form Q-20 during the evaluation process.

A chief concern of the inspection agency is the actual implementation of the QC plan itself. In other words, it is not only the fact that the manufacturer or fabricator *has* a plan that is important; it is the "effective implementation" of the plan that matters most. To quote the ICC-ES, "the inspector should always review both the manufacturer's quality documentation and the *effective implementation* of the quality system." The plan by itself will not ensure quality. Rather, it is how the manufacturer applies the plan that ensures a quality, reliable product.

Seek Green Products with an Evaluation Report

Manufacturers that have obtained approvals for their building products and systems know and understand the importance of the evaluation report and its relationship to a given construction project. Given the complexities associated with the approval process and the time and expense involved in obtaining a report, these manufacturers are eager to provide you with this information because they know it sets their products apart from the competition. Products carrying an approval through the ICC-ES are often readily accepted by the code official, so it is certainly wise to seek out green building products and systems for which the manufacturer has obtained an evaluation report.

The code official may also accept a product approval report through a legacy organization such as the International Conference of Building Officials (ICBO), Southern Building Code Congress International (SBCCI), Building Officials and Code Administrators (BOCA), or the National Evaluation Service (NES). Legacy reports are product approval reports that were generated prior to the formation of the International Code Council Evaluation Service in February 2003 when the nation's four building product evaluation services combined into one organization. Many of these reports are still in use today.

However, not all building products or systems will be approved under the ICC-ES system. The manufacturer may not be aware of the product evaluation process or may not have considered the advantages and many benefits of product approval. Also, since the product testing and approval process can be time consuming and expensive, some manufacturers may be reluctant to pursue such approvals. In addition, for building systems such as earth-sheltered construction, straw bale, and similar, there are no standards or tests by which an evaluation can be made. When working with the manufacturer of a green product, encourage the maker to seek product approval through the ICC-ES or other approval agency where possible as it may save time and money in the long run. Remember, however, that the lack of an evaluation report does not necessarily mean that a product or system is inferior, nor should it be assumed that a report is necessary to obtain approval from the code official.

CONSTRUCTION METHODS AND THE IRC

Like construction materials, construction methods are also regulated under the IRC. Once a particular material has been chosen for use, the code establishes specific requirements for the way the materials are to be assembled, how they are fastened, and what limitations, if any, apply. The code, then, provides an instruction manual of sorts that *prescribes* how the construction is to be done.

In code parlance, the term **prescriptive construction** is often used to refer to construction methods that comply with the basic requirements of the code. These methods are based on traditional forms of construction with long-established histories of success and durability. In simple terms, the code uses a "cookbook" approach to prescribe to the reader the manner in which a particular form of construction should take place.

The IRC establishes code requirements for construction methods on a variety of levels. Requirements for the basic structure of a dwelling (foundations, walls, floors, roofs) are spelled out in great detail in Chapters 4 through 10 of the IRC. It establishes criteria in other areas as well. For example, Chapter 3 of the IRC, Building Planning, provides basic code requirements for a variety of safety-related items. Here you will find requirements for safety glazing, guards, stairways, emergency egress, and a host of other safety-related code provisions (see Figure 2-7).

Figure 2-7

Code-prescribed framing for stairs. Note stair width, riser height, and tread width, all of which are prescribed by the International Residential Code.

COURTESY DELMAR/CENGAGE LEARNING

Chapter 11, Energy Conservation, provides code provisions to ensure dwellings are energy efficient. The following will explore in detail some basic ways in which construction methods are regulated.

Prescriptive Wood Framing Requirements

In a previous section, we explored how sawn lumber is regulated as a construction material. Recall that sawn lumber is typically graded and identified by an indelible stamp. Also recall that the structural properties for a particular piece of sawn lumber vary by species and grade, which largely determine its load-carrying capacity.

So how exactly does the IRC use this information to prescribe a specific construction method? Through the written text and through the use of tables or figures, the code spells out how a particular piece of lumber can be used in the structure. Take wall studs for example. The provisions for walls are found in Chapter 6 of the IRC. Section R602 prescribes how walls are to be constructed, including code provisions for stud size, height and spacing, top plate configuration, provisions for notching and drilling, and so forth. These provisions are based on the assumed minimum lumber grades already established in the previous section (No. 3, standard or stud grade).

In many cases, this information is placed in tables for ease of use, such as the one shown in Figure 2-8. Table 602.3(5) conveniently provides information on stud size, spacing, and height limitations for both bearing and nonbearing walls. Note that as stud size increases and/or the spacing between studs decreases, the load carrying capacity for the wall increases. In some cases, the use of a particular stud size is prohibited in a particular wall configuration. In these cases, the horizontal line in that column indicates that the particular stud size is prohibited, as is the case for 2 inch by 3 inch (51 mm by 76 mm) studs in bearing walls. It is a relatively simple matter, then, to determine the stud size, height, and spacing requirements.

Similar tables, known as **span tables**, are provided for floor and roof framing as well. Once the species and grade of the lumber is known, tables contained in the code provide allowable spans for a variety of loading conditions. For example, the IRC establishes a live load requirement of 30 pounds per square foot (psf) (1.437 kPa) in sleeping rooms. Look at Table R502.3.1(1) in Figure 2-9. This table provides allowable spans for sleeping rooms with a live load of 30 psf (1.437 kPa), so we can use it to determine floor joist spans in a bedroom. If, in this example, we choose a joist spacing of 12 inches (305 mm) and also assume a lumber species

Figure 2-8

Table R602.3(5) Size, height, and spacing of wood studs.

TABLE R602.3(5)
SIZE, HEIGHT AND SPACING OF WOOD STUDS[a]

| STUD SIZE (inches) | BEARING WALLS | | | | | NONBEARING WALLS | |
	Laterally unsupported stud height[a] (feet)	Maximum spacing when supporting a roof-ceiling assembly or a habitable attic assembly, only (inches)	Maximum spacing when supporting one floor, plus a roof-ceiling assembly or a habitable attic assembly (inches)	Maximum spacing when supporting two floors, plus a roof-ceiling assembly or a habitable attic assembly (inches)	Maximum spacing when supporting one floor height[a] (feet)	Laterally unsupported stud height[a] (feet)	Maximum spacing (inches)
2 × 3[b]	—	—	—	—	—	10	16
2 × 4	10	24[c]	16[c]	—	24	14	24
3 × 4	10	24	24	16	24	14	24
2 × 5	10	24	24	—	24	16	24
2 × 6	10	24	24	16	24	20	24

For SI: 1 inch = 25.4 mm, 1 foot = 304.8 mm, 1 square foot = 0.093 m².

a. Listed heights are distances between points of lateral support placed perpendicular to the plane of the wall. Increases in unsupported height are permitted where justified by analysis.

b. Shall not be used in exterior walls.

c. A habitable attic assembly supported by 2 × 4 studs is limited to a roof span of 32 feet. Where the roof span exceeds 32 feet, the wall studs shall be increased to 2 × 6 or the studs shall be designed in accordance with accepted engineering practice.

and grade of Douglas fir-larch, No. 2, we see that a 2 inch by 10 inch (51 mm by 254 mm) joist will span 19 feet 10 inches (6,045 mm) with a 10 psf (0.479 kPa) dead load. A similar procedure is used for rafter and ceiling joist spans, the provisions for which can be found in Chapter 8 of the IRC.

The IRC prescribes methods for wood construction similarly throughout the code. Thus, where wood is used, the IRC establishes basic code requirements and provides extensive information to assist in the construction of a dwelling.

Prescriptive Requirements for Concrete Foundations

As with wood framing, the IRC also provides prescriptive requirements for foundations. Provisions for wood, concrete, and masonry foundations are found in Chapter 4, along with a number of tables and illustrations designed to assist in the construction. As with wood, the IRC prescribes minimum compressive strength for concrete, minimum soil bearing values, footing widths and thicknesses, foundation wall widths and thicknesses, as well as requirements for **rebar** (reinforcing steel), anchor bolt size and spacing, and more.

Again for ease of use, the IRC conveniently tabulates much of this information to assist the reader in determining minimum code requirements. In a previous section, it was established that the minimum compressive strength for most conditions is 2,500 psi (17,237 kPa), which can easily be seen in Table R402.2

Figure 2-9 Table R502.3.1(1) Floor joist spans for common lumber species.

TABLE R502.3.1(1)
FLOOR JOIST SPANS FOR COMMON LUMBER SPECIES
(Residential sleeping areas, live load = 30 psf, L/Δ = 360)[a]

JOIST SPACING (inches)	SPECIES AND GRADE		DEAD LOAD = 10 psf				DEAD LOAD = 20 psf			
			2 × 6	2 × 8	2 × 10	2 × 12	2 × 6	2 × 8	2 × 10	2 × 12
			Maximum floor joist spans							
			(ft - in.)	(ft - in.)	(ft - in.)	(ft - in.)	(ft - in.)	(ft - in.)	(ft - in.)	(ft - in.)
12	Douglas fir-larch	SS	12-6	16-6	21-0	25-7	12-6	16-6	21-0	25-7
	Douglas fir-larch	#1	12-0	15-10	20-3	24-8	12-0	15-7	19-0	22-0
	Douglas fir-larch	#2	11-10	15-7	19-10	23-0	11-6	14-7	17-9	20-7
	Douglas fir-larch	#3	9-8	12-4	15-0	17-5	8-8	11-0	13-5	15-7
	Hem-fir	SS	11-10	15-7	19-10	24-2	11-10	15-7	19-10	24-2
	Hem-fir	#1	11-7	15-3	19-5	23-7	11-7	15-2	18-6	21-6
	Hem-fir	#2	11-0	14-6	18-6	22-6	11-0	14-4	17-6	20-4
	Hem-fir	#3	9-8	12-4	15-0	17-5	8-8	11-0	13-5	15-7
	Southern pine	SS	12-3	16-2	20-8	25-1	12-3	16-2	20-8	25-1
	Southern pine	#1	12-0	15-10	20-3	24-8	12-0	15-10	20-3	24-8
	Southern pine	#2	11-10	15-7	19-10	24-2	11-10	15-7	18-7	21-9
	Southern pine	#3	10-5	13-3	15-8	18-8	9-4	11-11	14-0	16-8
	Spruce-pine-fir	SS	11-7	15-3	19-5	23-7	11-7	15-3	19-5	23-7
	Spruce-pine-fir	#1	11-3	14-11	19-0	23-0	11-3	14-7	17-9	20-7
	Spruce-pine-fir	#2	11-3	14-11	19-0	23-0	11-3	14-7	17-9	20-7
	Spruce-pine-fir	#3	9-8	12-4	15-0	17-5	8-8	11-0	13-5	15-7
16	Douglas fir-larch	SS	11-4	15-0	19-1	23-3	11-4	15-0	19-1	23-0
	Douglas fir-larch	#1	10-11	14-5	18-5	21-4	10-8	13-6	16-5	19-1
	Douglas fir-larch	#2	10-9	14-1	17-2	19-11	9-11	12-7	15-5	17-10
	Douglas fir-larch	#3	8-5	10-8	13-0	15-1	7-6	9-6	11-8	13-6
	Hem-fir	SS	10-9	14-2	18-0	21-11	10-9	14-2	18-0	21-11
	Hem-fir	#1	10-6	13-10	17-8	20-9	10-4	13-1	16-0	18-7
	Hem-fir	#2	10-0	13-2	16-10	19-8	9-10	12-5	15-2	17-7
	Hem-fir	#3	8-5	10-8	13-0	15-1	7-6	9-6	11-8	13-6
	Southern pine	SS	11-2	14-8	18-9	22-10	11-2	14-8	18-9	22-10
	Southern pine	#1	10-11	14-5	18-5	22-5	10-11	14-5	17-11	21-4
	Southern pine	#2	10-9	14-2	18-0	21-1	10-5	13-6	16-1	18-10
	Southern pine	#3	9-0	11-6	13-7	16-2	8-1	10-3	12-2	14-6
	Spruce-pine-fir	SS	10-6	13-10	17-8	21-6	10-6	13-10	17-8	21-4
	Spruce-pine-fir	#1	10-3	13-6	17-2	19-11	9-11	12-7	15-5	17-10
	Spruce-pine-fir	#2	10-3	13-6	17-2	19-11	9-11	12-7	15-5	17-10
	Spruce-pine-fir	#3	8-5	10-8	13-0	15-1	7-6	9-6	11-8	13-6
19.2	Douglas fir-larch	SS	10-8	14-1	18-0	21-10	10-8	14-1	18-0	21-0
	Douglas fir-larch	#1	10-4	13-7	16-9	19-6	9-8	12-4	15-0	17-5
	Douglas fir-larch	#2	10-1	12-10	15-8	18-3	9-1	11-6	14-1	16-3
	Douglas fir-larch	#3	7-8	9-9	11-10	13-9	6-10	8-8	10-7	12-4
	Hem-fir	SS	10-1	13-4	17-0	20-8	10-1	13-4	17-0	20-7
	Hem-fir	#1	9-10	13-0	16-4	19-0	9-6	12-0	14-8	17-0
	Hem-fir	#2	9-5	12-5	15-6	17-1	8-11	11-4	13-10	16-1
	Hem-fir	#3	7-8	9-9	11-10	13-9	6-10	8-8	10-7	12-4
	Southern pine	SS	10-6	13-10	17-8	21-6	10-6	13-10	17-8	21-6
	Southern pine	#1	10-4	13-7	17-4	21-1	10-4	13-7	16-4	19-6
	Southern pine	#2	10-1	13-4	16-5	19-3	9-6	12-4	14-8	17-2
	Southern pine	#3	8-3	10-6	12-5	14-9	7-4	9-5	11-1	13-2
	Spruce-pine-fir	SS	9-10	13-0	16-7	20-2	9-10	13-0	16-7	19-6
	Spruce-pine-fir	#1	9-8	12-9	15-8	18-3	9-1	11-6	14-1	16-3
	Spruce-pine-fir	#2	9-8	12-9	15-8	18-3	9-1	11-6	14-1	16-3
	Spruce-pine-fir	#3	7-8	9-9	11-10	13-9	6-10	8-8	10-7	12-4
24	Douglas fir-larch	SS	9-11	13-1	16-8	20-3	9-11	13-1	16-2	18-9
	Douglas fir-larch	#1	9-7	12-4	15-0	17-5	8-8	11-0	13-5	15-7
	Douglas fir-larch	#2	9-1	11-6	14-1	16-3	8-1	10-3	12-7	14-7
	Douglas fir-larch	#3	6-10	8-8	10-7	12-4	6-2	7-9	9-6	11-0
	Hem-fir	SS	9-4	12-4	15-9	19-2	9-4	12-4	15-9	18-5
	Hem-fir	#1	9-2	12-0	14-8	17-0	8-6	10-9	13-1	15-2
	Hem-fir	#2	8-9	11-4	13-10	16-1	8-0	10-2	12-5	14-4
	Hem-fir	#3	6-10	8-8	10-7	12-4	6-2	7-9	9-6	11-0
	Southern pine	SS	9-9	12-10	16-5	19-11	9-9	12-10	16-5	19-11
	Southern pine	#1	9-7	12-7	16-1	19-6	9-7	12-4	14-7	17-5
	Southern pine	#2	9-4	12-4	14-8	17-2	8-6	11-0	13-1	15-5
	Southern pine	#3	7-4	9-5	11-1	13-2	6-7	8-5	9-11	11-10
	Spruce-pine-fir	SS	9-2	12-1	15-5	18-9	9-2	12-1	15-0	17-5
	Spruce-pine-fir	#1	8-11	11-6	14-1	16-3	8-1	10-3	12-7	14-7
	Spruce-pine-fir	#2	8-11	11-6	14-1	16-3	8-1	10-3	12-7	14-7
	Spruce-pine-fir	#3	6-10	8-8	10-7	12-4	6-2	7-9	9-6	11-0

For SI: 1 inch = 25.4 mm, 1 foot = 304.8 mm, 1 pound per square foot = 0.0479 kPa.

Note: Check sources for availability of lumber in lengths greater than 20 feet.

a. Dead load limits for townhouses in Seismic Design Category C and all structures in Seismic Design Categories D_0, D_1 and D_2 shall be determined in accordan with Section R301.2.2.2.1.

Figure 2-10

Table R402.2 Minimum specified compressive strength of concrete.

TABLE R402.2
MINIMUM SPECIFIED COMPRESSIVE STRENGTH OF CONCRETE

TYPE OR LOCATION OF CONCRETE CONSTRUCTION	MINIMUM SPECIFIED COMPRESSIVE STRENGTH[a] (f'_c)		
	Weathering Potential[b]		
	Negligible	Moderate	Severe
Basement walls, foundations and other concrete not exposed to the weather	2,500	2,500	2,500[c]
Basement slabs and interior slabs on grade, except garage floor slabs	2,500	2,500	2,500[c]
Basement walls, foundation walls, exterior walls and other vertical concrete work exposed to the weather	2,500	3,000[d]	3,000[d]
Porches, carport slabs and steps exposed to the weather, and garage floor slabs	2,500	3,000[d, e, f]	3,500[d, e, f]

For SI: 1 pound per square inch = 6.895 kPa.
a. Strength at 28 days psi.
b. See Table R301.2(1) for weathering potential.
c. Concrete in these locations that may be subject to freezing and thawing during construction shall be air-entrained concrete in accordance with Footnote d.
d. Concrete shall be air-entrained. Total air content (percent by volume of concrete) shall be not less than 5 percent or more than 7 percent.
e. See Section R402.2 for maximum cementitious materials content.
f. For garage floors with a steel troweled finish, reduction of the total air content (percent by volume of concrete) to not less than 3 percent is permitted if the specified compressive strength of the concrete is increased to not less than 4,000 psi.

in Figure 2-10. This simple table shows that, depending on the location of the concrete construction and its potential for weathering, in most cases, the code-required minimum is in fact 2,500 psi (17,237 kPa).

In outdoor locations where foundation walls or slabs receive maximum exposure to the weather, in particular where weather is moderate to severe, the minimum compressive strength increases to as high as 3,500 psi (24,132 kPa). It is important to note that a number of footnotes appear below the table. As mentioned already, this is typical throughout the code. Pay particular attention to these notes when they appear, as they often provide critical information that clarifies and sometimes limits the data contained in the table.

Additionally, the IRC also provides helpful illustrations, commonly referred to as details, to further clarify the intent of the code provisions. Figure R403.1(1) shows a number of examples of concrete and masonry footing details (see Figure 2-11). Footing widths, W, and required projections, P, are shown where applicable and coincide with the text presented in Section R403.1.1. Another important note, however, needs to be made here. While the illustrations are useful in that they provide a graphical representation of the code language, it is the text of the code section itself that presents the actual code requirements. The illustrations are provided as a tool to assist the reader. Where discrepancies occur, the code text takes precedence.

LOOKING AHEAD

Chapter 2 provided a detailed look at the way in which the IRC regulates construction materials and methods. With an understanding of the IRC's treatment of building materials and prescriptive construction methods as well as a keen

Figure 2-11

Figure R403.11(1) Concrete and masonry foundation details.

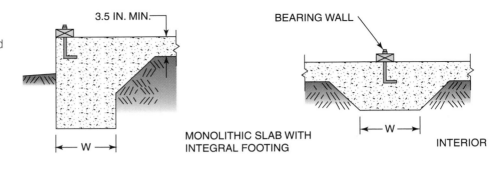

MONOLITHIC SLAB WITH INTEGRAL FOOTING

INTERIOR

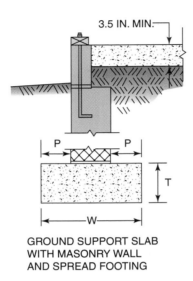

GROUND SUPPORT SLAB WITH MASONRY WALL AND SPREAD FOOTING

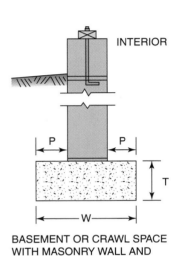

BASEMENT OR CRAWL SPACE WITH MASONRY WALL AND SPREAD FOOTING

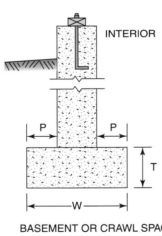

BASEMENT OR CRAWL SPACE WITH CONCRETE WALL AND SPREAD FOOTING

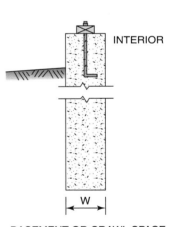

BASEMENT OR CRAWL SPACE WITH FOUNDATION WALL BEARING DIRECTLY ON SOIL

For SI = 1 inch = 25.4 mm

understanding of the product approval process, we now have the background necessary to advance our study of the code as it relates to green building. This will be useful as we explore areas where green options exist in the IRC. This knowledge, combined with what we learned in Chapter 1, prepares us to understand how to apply the code to green construction which we will explore fully in Chapter 3.

GREEN CONSTRUCTION MATERIALS, METHODS, AND THE IRC

learning objective

To know and understand how green materials and methods are treated under the provisions of the International Residential Code and to apply the principles learned in Chapter 2 to various forms of green construction.

THE INTERFACE BETWEEN GREEN AND THE IRC

In Chapter 2 we learned specifically how the IRC regulates construction materials and methods. Through the use of referenced standards, the code imposes standards for quality for all construction materials. We learned that product evaluation reports provide additional assurance to the code official that materials and systems are code compliant. We also learned how the IRC regulates construction methods through prescriptive code requirements. We are now ready to apply these concepts to green building in the next step toward going green with the IRC.

Green Construction Materials

Although it may come as a revelation, the IRC treats green building materials and methods in much the same way as it treats traditional or conventional forms of construction. This fact is quite remarkable, because most people expect that green building materials and methods of construction are regulated in an entirely different way than traditional materials. In fact, this expectation perpetuates a commonly held belief that the IRC actually prohibits green construction.

The code is, however, neutral on the subject of green building. It neither promotes nor prohibits green materials or construction methods. As a result, green materials must meet the same standards as conventional materials—a fact that is both good and bad. It is good because there are no additional or unusual code requirements for green products, creating essentially a level playing field for all materials. It is bad because it is often the case that standards for "green" focus on matters unrelated to the concerns of the code or, in some cases, are simply non-existent. Those green standards available today focus generally on the attributes that make a product sustainable (not currently a concern of the code), many of which are developing or changing at a rapid pace as the latest trends in green thinking emerge.

Green Attributes

Unlike building code standards that address issues such as flame spread, smoke development, and weather resistance, green standards focus on green attributes. That is, they focus on those characteristics that make the particular item an environmentally friendly, sustainable choice. These attributes often include considerations for indoor air quality, impact of the development and construction on the site, environmental impact in general, use of sustainable materials, and other related topics.

These attributes are very important, no doubt, but generally fall outside the concerns of the code. In the context of the traditional code-compliance plan review, the plan reviewer will likely never send you a letter saying that the building material you have chosen is not sustainable enough or will not meet standards for, say, indoor air quality because these things are not currently regulated by the code. It is essential, then, that one develop the ability to differentiate between those attributes that make a product green and those characteristics of the building product that are regulated by the code.

Testing

Some green building materials, such as structural insulated panels (SIPs), undergo extensive testing by the manufacturer. SIPs are "sandwich" panels made up of two layers of wood structural panels glued to a layer of foam plastic insulation (see Figure 3-1). Panels can be stood on edge to form walls, laid flat to form floors, or sloped to form roofs. As an engineered product, SIPs are consistent from panel to panel in terms of strength and quality. They can be worked with the same tools used for conventional construction and they are also highly energy efficient. SIPs have gained popularity in recent years and for this reason may already be familiar to you or the code official.

Tests on SIPs are performed by accredited labs to determine characteristics such as structural properties and load carrying capacities, flame spread and smoke-developed ratings, insulating capabilities and limitations, as well as a variety of other technical data useful in the design of a dwelling. Testing is performed to recognized standards in much the same way as conventional building materials. In the case of SIPs, it is likely that ICC-ES evaluation reports

Figure 3-1

An illustration of a structural insulated panel (SIP) showing a window and door opening. Note the outer skin, which is formed by structural sheathing, and the inner core, which is formed of structural foam.

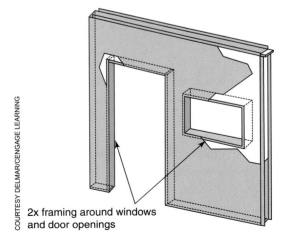

2x framing around windows and door openings

Measure It Twice...Cut It Once

As this old saying implies, it is better to proceed with care and caution than to regret one's actions later. To illustrate, I offer the following true story.

Several years ago, I was approached by one of my inspectors who expressed concern over the installation of a brand new spray-applied foam insulation product he encountered while performing his daily inspections. Neither he nor I had encountered a product of this type before. He found, upon inspection, that the foam insulation had been installed in both the crawl and attic spaces of a dwelling. The foam material was sprayed onto all surfaces within those spaces in varying thicknesses and appeared to actually be adhered to the exposed wood, concrete, and gypsum surfaces. Knowing that foam plastics are heavily regulated under the provisions of, what was then, the 2006 edition of the IRC and also seeing that the product was not specified on the approved construction plans, he correctly asked for documentation showing that the product was approved for use and was compliant with Section R314 Foam Plastic in the IRC.

The product itself—essentially a combination of insulating foam plastic and adhesive—had some testing but the manufacturer had not yet had the product evaluated by an approved evaluation agency. Thus, there was no independent assurance that the product would perform within the limitations imposed by the code. In the end, the test data and other documentation showed the product did not comply with the requirements of Section R314.3 (2006 edition of the IRC), which established a maximum flame spread index of 75 for foam plastics tested to 4 inches in thickness, nor had it been specifically tested at the installed thicknesses greater than 4 inches as required by Section 314.6 (2006 edition of the IRC).

It seemed that the contractor and homeowners jointly made the decision to purchase and install the product based on sales literature they had obtained at a home and garden show, rather than from actual test data and/or product approval reports. The product literature touted the super-insulating effects of the foam insulation and other benefits, such as the speed and ease of installation but gave little information regarding the testing methods used to evaluate the material or to what standards the material had been tested.

Unfortunately, the contractor and owners also installed the spray-applied foam insulation throughout the attic and crawl space of the dwelling without first checking to see if it was an approved product. They did not do so maliciously; rather, they were caught up in the excitement of utilizing a product they believed to be superior to conventional fiberglass batt insulation (a decidedly green thing to do). In fact, at that time, they were completely unaware that they should have checked first, as it had never entered their minds!

Even after exploring many solutions, including the possibility of adding an automatic fire sprinkler system to the attic and crawl or separating the foam plastic somehow from the enclosed spaces with a thermal barrier, no practical solution was found. As a result, *all* of the installed foam material had to be removed from the attic and crawl space areas. Because the product was actually adhered to the floor and roof surfaces, removal also required *scraping* the product off of every surface in the attic and crawl space and disposing of it, which was a messy, tedious, and time-consuming job.

This expensive and labor-intensive tear-out could have been avoided if the building owner and contractor had simply checked with the building department prior to purchasing and installing the product—a point I have made repeatedly in this text. Be sure to thoroughly investigate all building products before you have made a commitment to use them in your green dwelling project. Errors such as the one described here can easily be avoided with some careful planning and communication.

or legacy reports are available, making the job of demonstrating compliance with the code rather easy.

In many cases, however, green building materials have limited or no testing. In these cases, it will be difficult to demonstrate to the code official how the material complies with the provisions of the code. Without test data showing the flame spread rating for the product, for example, the code official may not accept the material for use because he cannot be certain that the product, if ignited, will not burn too rapidly.

If you are a homeowner considering the use of a particular green product, ask the manufacturer if its products are tested and to what standards before you buy them to avoid costly mistakes. If you are a builder, discuss the products you want to install with the code official before you purchase and install them to avoid expensive tear-outs and delays. If you are a design professional, check for test data and product approval reports before you specify a green product to avoid delays during the code review process. The time you invest researching and verifying these things ahead is time well spent.

STANDARDS FOR GREEN CONSTRUCTION

If the IRC does not provide standards for green building, then who does? There are currently a number of standards-writing organizations for green construction, and the list is growing. These organizations are comprised of leading experts in the fields of green construction, environmental design, and conservation. The standards they produce have become or are likely to become widely accepted in the world of green building. As stated in the last section, these standards provide guidance and criteria as it relates to the green attributes of a building site, development, or building as a whole. Let's look at a few of these standards.

Leadership in Energy and Environmental Design for Homes

Figure 3-2
Leadership in Energy and Environmental Design (LEED) for Homes rating system insignia.

The U.S. Green Building Council's (USGBC) Leadership in Energy and Environmental Design (LEED) for Homes is a points-based standard designed to encourage the "homebuilding industry toward more sustainable practices." This USGBC standard is designed to work with all aspects of the homebuilding industry, by identifying and recognizing sustainable design and construction practices. A primary goal of the LEED for Homes standard is its desire to provide consistency in how green features are defined. LEED for Homes is a consensus standard developed by experts in the field of green building (see Figure 3-2).

LEED for Homes provides a points-based rating system in eight principle performance areas, some of which include Innovation and Design Process, Sustainable Sites, and Indoor Environmental

Quality. There are four performance levels designated as Certified, Silver, Gold, and Platinum. The LEED for Homes rating system establishes minimum mandatory requirements (known as prerequisites) and offers additional points for performance enhancements in each category.

The USGBC standard requires third-party verification through LEED for Homes providers and green raters, one of its great strengths. Through a network of carefully selected organizations, providers manage a team of green raters who, under contract with the USGBC recruit and register projects for LEED for Homes, coordinate and oversee the activities of green raters, certify LEED homes, provide quality assurance for the certifications, and coordinate with the USGBC and its chapters.

Green Home Building Guidelines

The National Association of Home Builders (NAHB) Green Home Building Guidelines (GHBG) is a comprehensive green building standard designed for the mainstream home builder. The NAHB recognizes that many builders today already utilize green building practices of one form or another. The GHBG provides additional guidance for builders to effectively implement green building practices in a holistic way, in an effort to minimize impact to the environment (see Figure 3-3).

Figure 3-3

Green Home Building Guidelines (GHBG) insignia.

COURTESY OF NAHB

Like the LEED for Homes standard, the GHBG is also a points-based system. At its core, the GHBG is developed around seven guiding principles, some of which include Resource Efficiency, Energy Efficiency, and Global Impact. The GHBG has three levels of performance: Bronze, Silver, and Gold. Levels are achieved by accumulating points in each of the seven guiding principle categories. Each level requires a minimum number of points, which must be earned in each of the seven scoring areas to ensure balance in the overall performance of the green home.

The GHBG standard offers a useful system for builders regardless of locale. By providing a comprehensive, ready-to-use standard, local associations with limited resources but who are still interested in a green building program can utilize the GHBG standard to great effect.

National Green Building Standard

The ICC-700-2008 National Green Building Standard (NGBS) is a recent publication produced by the International Code Council. Developed at the request of and in conjunction with the National Association of Home Builders, the NGBS rates site design, developments, and residential buildings for potential impact on the environment. Unlike other green standards, the NGBS is the first of its kind to be approved as an American National Standards Institute (ANSI) standard, making it a true consensus-developed option for the evaluation of green dwellings and developments (see Figure 3-4).

Figure 3-4

The National Green Building Standard (NGBS).

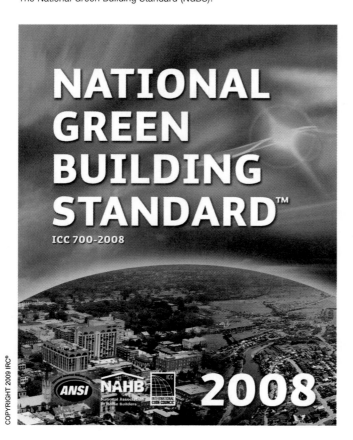

COPYRIGHT 2009 IRC®

The NGBS also rates residential buildings on a points system, like the previous two examples. The NGBS is a minimum standard and includes four designations: Bronze, Silver, Gold, and Emerald to encourage higher performance levels in green building. The NGBS also establishes four designations for sites and developments: One, Two, Three, and Four to encourage higher performance levels for sites.

Another unique feature of the NGBS is that it provides standards for existing buildings, which make up approximately 90% of all housing. The NGBS provides specific standards and ratings for additions and renovations, a major feature that sets it apart from most other green building programs.

A Word of Caution

It bears repeating here that the U.S. Green Building Council's LEED for Homes, the National Association of Home Builders' GHBG, and the International Code Council's NGBS are not a part of the IRC, nor are they referenced in any way in the actual code document. Thus, to the extent they provide standards, they cannot be interpreted as code requirements.

However, recall from an earlier chapter that many jurisdictions adopt local ordinances or implement zoning codes, many of which now contain incentives, and requirements, to build or develop in a green way. Readers are strongly cautioned to check with the local jurisdiction to determine if there are specific *local* requirements to build or develop "green" and to what standard, if any, construction must adhere. For example, the cities of Telluride and Boulder, Colorado, have mandatory requirements for compliance with green building ordinances and other criteria, independent of the requirements of the building code.

SUSTAINABLE ATTRIBUTES VERIFICATION AND EVALUATION PROGRAM

The ICC-ES offers a new service designed to provide information to designers, specifiers, and code officials about sustainable building products. The newly established Sustainable Attributes Verification and Evaluation (SAVE) Program is designed to provide objective evaluation criteria about claims made by the manufacturers of sustainable products. It is different from the green standards mentioned earlier in that, rather than focus on the way in which the dwelling or development on the whole impacts the environment, SAVE is focused on the

Figure 3-5

Products that are evaluated under the ICC-ES SAVE™ program and issued a Verification of Attributes Report™ (VAR™) are allowed to display this ICC-ES SAVE™ mark.

materials and processes that go into manufacturing the green product itself (see Figure 3-5).

A voluntary program, SAVE gives manufacturers the opportunity to have their products evaluated by the ICC-ES. The evaluation includes an inspection of the manufacturing facility and processes, along with analysis of testing in recognized laboratories. Evaluations are based on established guidelines for sustainability. The SAVE program allows those manufacturers seeking to provide accurate information about sustainable products a means to do so. A report produced under the SAVE program is known as a Verification of Attributes Report (VAR).

Manufacturers seeking product approval under the SAVE program have their products evaluated against guidelines developed with input from industry, the ICC, and other parties. The guidelines provide a basis by which manufacturers' claims about the sustainable attributes of their products can be evaluated. Attributes are verified in the following nine categories.

- Evaluation Guideline for Determination of Recycled Content of Materials (EG101)
- Evaluation Guideline for Determination of Biobased Material Content (EG102)
- Evaluation Guideline for Determination of Solar Reflectance, Thermal Emittance and Solar Reflective Index of Roof Covering Materials (EG103)
- Evaluation Guideline for Determination of Regionally Extracted, Harvested or Manufactured Materials or Products (EG104)
- Evaluation Guideline for Determination of Volatile Organic Compound (VOC) Content and Emissions of Adhesives and Sealants (EG105)
- Evaluation Guideline for Determination of Volatile Organic Compound (VOC) Content and Emissions of Paints and Coatings (EG106)
- Evaluation Guideline for Determination of Volatile Organic Compound (VOC) Content and Emissions of Floor Covering Products (EG107)
- Evaluation Guideline for Determination of Formaldehyde Emissions of Composite Wood and Engineered Wood Products (EG108)
- Evaluation Guideline for Determination of Certified Wood and Certified Wood Content in Products (EG109)

PLUMBING, MECHANICAL, AND FUEL GAS LISTING PROGRAM

Like the SAVE program, the ICC-ES also offers a new service designed to provide information to interested persons about products manufactured for use in the plumbing, mechanical, and fuel gas (PMG) industries. The newly established PMG program is designed to provide objective evaluation criteria that demonstrate that these products not only meet applicable standards but also comply with the requirements of applicable codes, such as the International Plumbing Code, the International Mechanical Code, and the International Fuel Gas Code, among others. It differs from other listing programs that typically only address compliance with applicable standards.

Also a voluntary program, the ICC-ES PMG listing service gives manufacturers of plumbing, mechanical, and fuel gas products the opportunity to have their products evaluated by the ICC-ES. The PMG listing evaluation includes a qualifying inspection of the manufacturing facility and processes, along with analysis of tests performed in recognized laboratories just like the ICC-ES's other evaluation services. PMG product evaluations are based on code-established, nationally recognized standards. In cases where there is no nationally recognized standard or where there is a nationally recognized standard but the code does not provide enough detail to address the permitted use of a particular product, the ICC-ES staff works closely with the product manufacturer to develop evaluation criteria. Reports produced under the PMG program are known as PMG Listings.

THE BUILDING CODE AS APPLIED TO GREEN CONSTRUCTION

Although the IRC does not address green building per se, there are fundamental elements contained within the green construction that it will always address. For example, regardless of whether traditional or green construction materials and methods are used, the code is always concerned with issues of structural stability, life safety, and energy efficiency. Let us explore some of the ways in which the IRC is applied to green construction.

Ecoroofs

Ecoroofs make a great case study in the application of the IRC, because roofs in general are regulated by relatively few sections of the code and are thus rather straightforward in terms of applying and understanding code requirements. Known by a variety of names, including green roofs or vegetated roofs, ecoroofs are rapidly increasing in popularity. In simplest terms, an **ecoroof** is a roof covered with various forms of vegetation (see Figures 3-6a and 3-6b).

Ecoroofs have many benefits. They reduce **heat island effect,** which is a growing issue and cause for concern in many densely populated areas. Heat island effect is the tendency for urban areas to retain heat from the sun during the day and then release it at night, creating warm zones and increasing average ambient temperatures. Ecoroofs offer an advantage in that they do not store heat; thus they stay cooler overall than traditional roofing materials.

They also produce oxygen due to the vegetation that is planted on the roofs, a welcome contribution in urban areas prone to smog and other types of pollution (see Figure 3-7). As with all plant matter, vegetation growing on an ecoroof also removes carbon dioxide from the air. Another benefit is that they absorb and utilize some rainwater, thereby reducing storm water runoff from roofs that can potentially harm streams. So, it is easy to see why they are growing in popularity and emerging in increasing numbers in residential construction.

Code Requirements

How exactly does the IRC regulate ecoroofs? Chapters 8 and 9 of the IRC address roof-ceiling construction and roof assemblies, respectively. Specific criteria are

Figure 3-6(a) & 3-6(b)

A sloped ecoroof installed on a small accessory building (Silverton, Oregon). Note the variety of plantings included in this design and the drainage holes along the eave.

(a)

(b)

provided in each chapter addressing the quality or grades of materials used; and prescriptive requirements are provided for the construction of roofs, ceilings, and the installation of roof coverings.

Chapter 8, Roof-Ceiling Construction, is concerned with the structure itself. As we have already learned, the IRC is concerned with roof framing and its ability to

Figure 3-7

Various forms of vegetation planted on an ecoroof (Silverton, Oregon). Succulent plants make an excellent, low maintenance choice for ecoroofs.

COURTESY DELMAR/CENGAGE LEARNING

Figure 3-8

Typical roof framing showing manufactured roof trusses, roof sheathing, and supporting elements.

COURTESY DELMAR/CENGAGE LEARNING

support roof live and dead loads (see Chapters 1 and 2 for a review if necessary). That is, the code provides specific criteria to ensure that roof trusses or rafters and ceiling joists are sized and spaced properly, that **roof sheathing** (roof plywood, oriented strand board, or particle board) is properly span rated and graded, and that related structural components are suitable for the loading conditions (see Figure 3-8).

Chapter 9, Roof Assemblies, is concerned with the specific roof coverings that will be placed on top of the roof sheathing and the means by which water is drained from the roof. These are the materials and systems that form a weather-resistant roof covering to protect the dwelling from water infiltration. In some applications, roof coverings must also meet requirements for fire and hail resistance. These materials include not only the roof covering (asphalt shingles, wood shakes, concrete tiles, etc.), but also include flashings, coping, and roof drains to prevent the intrusion of water.

The code provides specific criteria for all major forms of roof coverings to ensure proper treatment of the roof deck, appropriate minimum slope of the roof to ensure drainage, flashing requirements, fastening requirements to ensure that the roof covering stays put, and in some areas where climatic conditions dictate, requirements for ice barriers.

Even with its extensive coverage and treatment of roof framing and coverings, the code does not provide any criteria for the ecoroof itself. That is, there are no provisions that address types of plantings, depths of soil, method of irrigation, and the like. The "green" portion of the roof is outside the scope of the IRC and is thus unregulated by the provisions of the code. This presents an interesting dilemma, as some portions of the roof structure and roof coverings are regulated by the code (e.g., its load carrying abilities and weather resistance), but other portions are entirely outside the scope of the code as mentioned previously.

So what do we do? In this example, we see a marriage of both regulated and unregulated types of construction. Our approach to meeting the code requirements, then, simply follows the obvious. We will worry about those aspects of the code that are regulated and the remainder will not be an issue.

It stands to reason that the soil, water, and plantings will weigh more than the standard roofing materials. In fact, the dead loads from these roofs typically fall somewhere between 15 and 25 pounds per square foot over and above the dead loads for the conventional portions of the roof depending on soil depth and other factors. Thus, our ecoroof must be designed to consider both the standard dead loads (framing, sheathing, insulation, etc.) and the increased dead and live loads from the "eco" portion of the roof (soil, water, plants, etc.), the latter of which will fluctuate due to soil saturation, plant size, and other variables.

Recall also from Chapter 2 of going green with the IRC that the code provides maximum span tables to ensure that the structural portion of our roof is not overloaded. However, also recall that these span tables can be used only for prescriptive construction under standard loading conditions. In this case, due to the increased dead loads from the soil and plantings, the standard span tables provided in the code will not apply. Thus, the solution is to provide an engineering analysis where all of the dead and live loads have been considered in the design.

Similarly, we know from Chapter 9 of the IRC that roofs must be constructed in a manner that will prevent the intrusion of water and that all materials used in the construction of the roof must be suitable for the purpose intended. In addition, although the code provides prescriptive installation guidelines for many traditional forms of roofing materials, it does not address the planted portion of the roof.

The solution here is to cover the roof with code-compliant roofing materials specifically suited for the task, and then construct the ecoroof over it. In other words, the code-compliant roof is installed *beneath* the ecoroof, typically using a membrane or sealable form of roofing material. A number of these products are easily obtained and many are specifically approved for use in these types of application. Additionally, a number of these roofing products have evaluation reports available that may assist in the approval process.

Straw Bale

Another useful but perhaps more complex case study is in the use of straw bales to construct a dwelling. Although the use of straw in construction dates back for centuries in parts of Europe, in the United States the origin of straw bales appears to coincide with the invention of the mechanical baler in the late 1800s. As the name implies, bales of straw are used to form the actual walls of the dwelling rather than conventional 2 inch by 4 inch (51 mm by 102 mm) or 2 inch by 6 inch (51 mm by 152 mm) framing members (see Figure 3-9).

Figure 3-9

Straw bales used in the construction of a dwelling.

Straw offers many advantages, making it an excellent choice for green construction. Straw is a renewable, natural product. It is a by-product of the production of cereal grains and has a similar chemical makeup to that of wood. It grows in abundance in most regions and is readily available locally, an important consideration in terms of sustainability. Straw is an excellent insulator and when baled, the additional wall thicknesses provide added insulation and sound control. The use of straw also greatly reduces the amount of lumber used in the construction of a dwelling, which is important in areas where concerns exist around deforestation.

There are two primary types of straw bale construction: load bearing and nonload bearing or "in-fill." Load bearing straw bale construction uses the straw bales themselves as the structural support for the roof (see Figure 3-10). In some cases where codes regulating straw bales have been adopted, such as in Oregon, load bearing straw bales are limited to one story in height. Straw bales used in these applications are usually precompressed so that they do not settle excessively under the loads from the roof. Large amounts of settling can cause gaps in the roof construction and around windows or doors,

Figure 3-10

A load bearing straw bale wall. Note the unique top plate assembly consisting of a sheet of structural sheathing (i.e., plywood) with two 2 x 4 members fastened to each side. The roof framing will be supported directly on the top plate which is, in turn, supported by the straw bales.

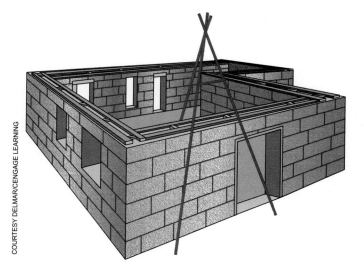

COURTESY DELMAR/CENGAGE LEARNING

Figure 3-11

A nonload bearing straw bale wall. In this case, the straw bales are used as in-fill between wood framing members.

COURTESY OF MARK PIEPKORN

which would eventually allow water into the structure (the enemy of straw bale), or at the least, create unsightly cracks in the exterior and interior plaster used to coat the walls and ceilings.

Nonload bearing straw bales, as the name implies, are applications where the bales are used as in-fill between structural frames that have been engineered to carry the weight of the structure and to resist live loads (see Figure 3-11). Nonload bearing straw bale structures are not typically limited to one story, because as in-fill, the bales are not carrying the structural loads. The structural frame can be easily designed and engineered to accommodate loads from multiple stories.

Code Requirements

The IRC does not regulate straw bale construction. However, as we have already seen there are elements contained within any dwelling that are addressed in the code. All structures, regardless of type, must be supported by a suitable foundation. They must be capable of safely resisting all applied loads (live loads due to snow, wind, seismic, etc.) and must meet all life safety requirements. They must meet requirements for energy conservation and contain building systems (mechanical, plumbing, and electrical) that meet all national standards for safety.

Straw bale construction falls generally into a category that is known as alternative materials and methods (AMM). This category is covered in detail in the next chapter, but for now, suffice it to say that it includes building materials and/or systems that fall entirely or partially outside the scope of the provisions of the IRC. Generally, any form of construction that falls outside of the prescriptive requirements of the code is regarded as an AMM.

At the time of this writing, there is no nationally recognized standard for straw bale construction. Some states and municipalities have developed their own prescriptive straw bale codes. The state of Oregon, for example, has adopted prescriptive provisions for both load bearing and nonload bearing straw bale construction in Appendix R of the Oregon Residential Specialty Code. The state of California; Pima County, Arizona; Austin, Texas; and Tucson, Arizona (among others) have also adopted codes and standards for straw bale.

In some locales, then, there are prescriptive codes to follow. In others, straw bale structures must be fully engineered and are regarded as AMM. In these cases, an engineered design must be provided to demonstrate to the satisfaction of the code official that the structure can safely support all loads and meet all applicable codes. In either case, as in our example for ecoroofs, those other elements of the straw bale dwelling that are regulated by the code must always be fully addressed, such as exiting and stairway requirements.

LOOKING AHEAD

Chapter 3 explored the ways in which the code does and does not regulate green construction. We learned how standards for green construction and development differ from code-adopted standards. We learned about several standards for sustainable development and we applied the code to two different types of green construction. We are now ready to explore alternative materials and methods, which we will do in detail in Chapter 4. We have nearly completed our preparatory work and are now more fully prepared in our quest toward going green with the IRC.

UNDERSTANDING ALTERNATIVE MATERIALS AND METHODS

learning objective

To know and understand what is meant by the IRC term "alternative materials and methods (AMM)" and how these underutilized code provisions can be usefully applied to a green construction project.

INTRODUCTION TO AMM

Leaf through any chapter in the IRC and it would be easy to reach the conclusion that only conventional construction (e.g., traditional wood framing, masonry, concrete) is allowed by the code. After all, virtually every code section, illustration, and table presents information about conventional construction and "prescribes" the course of action expected during the traditional construction process.

Remember, however, from our discussion in Chapter 1 that the IRC only establishes standards for construction for conventional or "prescriptive" construction. The intent, of course, is to provide a comprehensive, stand-alone code that utilizes time-tested, proven construction techniques and to minimize the need for engineering. Using a "cookbook" approach, the code simply spells out what one should or should not do in an effort to comply with the code.

Despite this focus on conventional methods of construction, however, the IRC also provides a means to deal with other-than-conventional methods of construction. Section R104.11 of the IRC addresses what are referred to in the world of building codes as "alternative materials, design, and methods of construction and equipment." The language contained therein represents, in my opinion, some of the most useful yet underutilized provisions in the IRC. This section states that it is not the intent of the code "to prevent the installation of any material or to prohibit any design or method of construction not specifically prescribed by this code, provided that any such alternative is approved" (see Figure 4-1). In this context, the word "approved" means acceptable to the code official.

As we see in the reference to Section R104.11 above, it is not the intent of the code to prohibit the use of new materials and new construction methods as they

Figure 4-1
A dwelling utilizing an alternative form of construction (straw bales). In this case, the structure is a hybrid, consisting of straw bales for the exterior walls and conventional wood frame construction at what appears to be a covered patio.

COURTESY OF SOLARWISE LLC—NATURAL BUILDING & DESIGN

are developed. In fact, one of the great strengths of the IRC is that it allows for other-than-conventional construction to be used so that, over time, as new technologies emerge and as innovative building materials are brought to market they can be used, assuming of course it can be determined that they satisfy all of the provisions of the code. Similarly, R104.11 also allows "traditional" materials and methods (e.g., straw and earth) to be used as alternates to the basic code requirements.

This brief but important section offers a powerful tool in the quest toward a green construction project, for two good reasons. First, it does not outright prohibit the use of green materials. Second and more important, it gives green building materials, designs, and construction methods equal footing under the code. In other words, the IRC does not impose any more restriction on green materials and methods than it would for traditional materials.

Many green building materials, systems, and construction methods fall into the category of AMM because they are not specifically prescribed in the body of the code. In fact, many green building systems, such as straw bale, structural insulated panels (SIPs) used as floor and roof panels, rammed earth, and others are not addressed in the main provisions of the code (see Figure 4-2). Our challenge is to demonstrate to the satisfaction of the code official that our proposed green building method is acceptable under the code.

The following sections will explore AMM in depth in an effort to provide the homeowner, builder, or architect with a deep understanding of how the IRC regulates AMM and how the provisions of these code sections can be used to seek approval of the proposed forms of green construction. Additionally, the 2009 *Performance Code for Buildings and Facilities* published by the ICC may also provide useful information for those interested in a more comprehensive discussion on the subject.

Figure 4-2
A steel floor joist fastened to insulating concrete forms (ICF) using a steel channel as a ledger. This installation is quite a departure from the typical wood-and-concrete systems used in conventional construction and is not directly addressed in the code.

COURTESY OF MARK MECKLEM, MIRANDA HOMES

APPROVAL OF AMM

There are two primary considerations in the approval of an AMM. Section R104.11 requires that an AMM be approved if the building official finds that the "proposed design is satisfactory and complies with the *intent* of the provisions of the code" and that the AMM is "for the purpose intended, at least the *equivalent* of that prescribed in the code." The following sections will explore each of these considerations more fully.

Intent of the Code

Consideration of the *intent* of the code is one of the great privileges enjoyed by the code official. It is through this code language that the code official is given the freedom to consider much more than the words printed on the page. Rather than being bound by the specific, literal language in a particular code section, she can consider what each particular code provision intends to do. That is, she can consider the intent *behind* the code language.

This consideration is important because it opens the door to the approval process for green building materials and methods, *if* it can be demonstrated that the proposed green construction meets the intent of the code. This is essential because, as we have already established, a great number of green building materials and construction methods do not meet the prescriptive provisions of the code (see Figure 4-3). If the code official were bound by the literal language of the code, it would be impossible to obtain approval for many of the green building materials and systems in use today.

Figure 4-3
A lovely example of rammed earth wall construction. Note the use of the massive rammed earth walls for this interior space, creating a beautiful setting for this dining room.

How to Determine the "Intent" of a Specific Code Provision

On the surface, determining the intent of the code seems straightforward. One would think that by simply reading the text, the intent of each code section would be apparent, and in some cases, it is. For example, Section R312 of the IRC requires guards around elevated floors, decks, and at the open sides of stairs. In this case, the intent of the code is clear: to prevent falls that could injure or kill someone. In other areas, however, the intent of the code is not so clear. There have been many instances in my career where someone *thought* she knew and understood the requirements of a particular code section only to discover that my interpretation varied greatly from her own. The point is that the true intent of the code is not always immediately apparent. Also, remember from our earlier discussion that it is the building official who is given the privilege of interpreting the code. This means that she gets to determine what is intended by each provision in the code. Seek to know and understand how the building official interprets the code and more importantly, its intent by talking with her at the earliest stages of the project.

Another tool that may be helpful in determining the intent of the code is a two-volume series called the 2006 *International Residential Code (IRC) Commentary* (the 2006 edition is the most recent available at the time of this writing). These two volumes provide analysis of many of the code's sections with helpful commentary. Over the years, they have proven extremely useful in determining the intent of the code because they provide background and sometimes a history of each code provision that can be used to better understand the language of the code. Remember, however, that the building official having authority in the jurisdiction where the green project will be built has the final say.

How to Determine "Equivalence" to the Code

It is helpful to think of the notion of *equivalence* by describing it with characteristics such as "quality, strength, effectiveness, fire-resistance, durability and safety." These words are borrowed from the Oregon Residential Specialty Code (Oregon's statewide amended version of the IRC) which does an excellent job of qualifying this important term. Additionally, the terms *weather resistance* and *energy efficiency* come to mind, which help our efforts to understand equivalence as it is used in the IRC.

These descriptors provide the reader a picture of the true breadth and depth of the code language. While this borrowed language is not contained in the original version of the IRC, it is helpful here because it illustrates rather clearly that the IRC is not just concerned with only one or two aspects of the code. Instead, the AMM must concern itself with *all* aspects of the code. Stated another way, an AMM must be demonstrated to be equivalent to that prescribed by the code in every way, not just in, say, the areas of energy efficiency or fire resistance.

This presents an interesting challenge for those who wish to build green. Often, manufacturers focus on only one or two areas of the code in their product literature. They may boast that their product is particularly weather-resistant or unusually durable, but leave out important information relevant to other parts of the code.

Remember, whether you are a homeowner, builder, or architect, you are trying to present an AMM that meets *all* of the requirements of the code, not just a few. The old but useful warning *caveat emptor* (let the buyer beware) is appropriate here. Do not rely solely on sales literature when evaluating a particular product, as this type of literature will often not provide the complete picture. Ask for test data and proof that a product or system has been thoroughly tested. Ask how it complies with specific standards and code requirements. These are sometimes hard questions for the manufacturer or supplier, but answers to them are essential if the AMM proposal has any hope of approval. These are the very questions the code official will ask, so it is wise to have answers when they ultimately arise.

While the preceding list is comprehensive, it is by no means meant to imply that these are the only concerns of the code official. When it comes to green building materials and systems, the code official may identify other areas of concern as well and each must be addressed in a thorough and meaningful way if she is to ultimately approve your proposal. The message here is simply this: Talk to her early and often so that the requirements of the jurisdiction are understood.

Figure 4-4

A residential fire sprinkler system installed in a beautiful, finished kitchen. Modern sprinkler systems are not only extremely effective, but also practically invisible when installed.

TRADE-OFF APPROACH

No discussion of AMM would be complete without an exploration of the trade-off approach. As the name implies, a trade-off approach occurs when certain **code enhancements** (i.e., the addition of certain safety features that would not otherwise be required) are made in exchange for the allowance of the AMM. For example, one might offer to add an automatic fire sprinkler system throughout the structure in exchange for allowing the use of, say, a new green insulation product with minimal testing (see Figure 4-4). In effect, a significant improvement to life safety (the addition of the fire sprinkler system) is provided in exchange for the increased "risk" associated with the approval of the AMM.

There are a few important things to note here. First, it is important to under-stand that the "trade" is only an enhancement to the overall safety of the project if the proposed life safety system or construction method is not required by the code in the first place. In a case where the IRC already requires the instal-lation of the fire sprinkler system, there would be no "value added" in offering to provide one because there is no choice but to provide it to comply with the requirements of the code. On the other hand, the code official may be satisfied that the *required* fire sprinkler system offers enough protection outright, espe-cially given the excellent reputation for fire protection and proven track record such systems typically provide. If so, the AMM proposal could be accepted on that basis alone.

Beginning with 2009 edition of the IRC, an automatic fire sprinkler system meet-ing the requirements of the NFPA Standard 13-D is required in all residences per-mitted for construction after the date the 2009 code is adopted. It is important to note here that because the NFPA 13-D fire sprinkler system will be a required system in all new construction, the use of it as a trade-off for other AMM consid-erations may, alas, no longer be an option in some jurisdictions. This is the first time in history that such a requirement is in place for residential structures, but its incorporation into the code illustrates what has been known for many years: Auto-matic fire sprinklers are highly effective at protecting the occupants of structures from the ravages of fires.

Does this mean, then, there are no longer any useful options available for use in a trade-off approach? No, of course not! Recall from an earlier chapter that the code is frequently amended at a local and sometimes even at a state level. It is entirely possible that a state or local jurisdiction could eliminate the new sprin-kler requirement through its own code adoption process. If this is the case, an automatic fire sprinkler system meeting the requirements of NFPA 13-D may still be considered a viable option in this approach to AMM. Be sure to check with your local building department for complete information.

Additionally, other trade-off options exist. For example, assuming your state or local jurisdiction retains the requirement for fire sprinklers, the following may prove useful as code enhancements.

Fire-Resistive Construction

Residential construction is typically *nonrated*. That is, there are no special fire-resistive protections provided for the structural components of the building, such as the roof framing, bearing walls, and floor systems. Fire-resistive construction, on the other hand, provides special protection for the significant structural mem-bers in a building, often through the use of certain gypsum board or by using mate-rials that are naturally fire resistant, such as masonry or concrete.

Fire-resistive construction offers a significant enhancement to life safety, because the main structural components provide an extra measure of protection from fire. This additional protection might be useful in a trade-off because, through the use

Figure 4-5

Fire-rated gypsum wallboard used to protect wood framing at a wall separating two dwelling units. Note the label identifying the product as fire resistive.

COURTESY DELMAR/CENGAGE LEARNING

of additional layers of fire-resistant gypsum board, concrete, masonry and other materials, the spread of fire is slowed considerably, allowing more time for building occupants to escape and more time for the fire department to respond as well (see Figure 4-5).

Combinations of some conventional and nonconventional materials might also be effective at increasing fire resistance. For example, cement-lime plaster, when applied to straw bale construction, has been demonstrated to meet a 2-hour fire rating through passage of the stringent ASTM E119 2-hour fire test with hose stream—more than twice the fire-resistive rating of conventional wall frame construction. In addition, the ASTM E84 test used to determine surface burning characteristics yielded a flame spread index of 10 and a smoke-developed index of 350 for this same assembly—impressive numbers to be sure!

Smoke Detection and Alarms

As we saw previously, the IRC requires a typical residence to be provided with smoke alarms (also called smoke detectors) in each bedroom, outside each sleeping area, and on each floor level. These alarms are typically the residential type and are usually not monitored by an approved monitoring agency. Since these alarms are required by the code, it would appear that there is no way to use them as a trade-off for an AMM.

There are, however, a number of commercially available smoke detection/alarm systems (like those seen in commercial buildings) that can be installed throughout a residence and offer significantly improved life safety through earlier detection and warning. Components used in these systems are considerably more sophisticated than their residential counterparts. These components are "smart" in that they do much more than simply detect smoke. For example, rate-of-rise detectors can be highly effective in certain areas of the dwelling. Rather than detect smoke, rate-of-rise detectors sense how quickly temperatures change at the ceiling. Under fire conditions, temperature rises very quickly, which is sensed by the rate-of-rise detector and thus trips an alarm. These are particularly useful near kitchens where ordinary smoke detectors often trip unnecessarily and where the smoke alarm manufacturer may actually prohibit their installation.

A Yard by Any Other Name…

As the following real-world example illustrates, a little thought and creativity can generate a solution to a code problem in a way that is not always immediately apparent. Recall that at the heart of any AMM is the need to meet the intent of the IRC's provisions. But this does not always require *adding* fire safety systems or providing fire-resistive construction as discussed in the previous sections. The following AMM demonstrates an equally effective method for achieving code compliance without the need for specialized construction methods and one I have used successfully many times over the course of my career.

Shortly after I became a building official, I was approached by an architect who was seeking a solution for a problem she was experiencing in the design of a dwelling. Her clients had purchased a small residential lot with a number of challenges, including a steeply sloped, and thus unusable, side yard similar to other surrounding properties. In order to make the project feasible, it was necessary to place the house directly on the opposite property line. These design limitations required the dwelling design to be quite narrow, making it impossible to avoid placing some of the bedrooms along the wall on the property line. Unfortunately, the IRC prohibits openings in walls constructed on a property line, creating a seemingly impossible situation in which the emergency escape windows required by Section R310.1 for the bedrooms could not be provided as required by the code.

Section R302.1 and Table R302.1 of the IRC establish safe clearances from property lines for unprotected exterior walls and also establish conditions under which additional fire protection is required for walls, projections, and other elements. These code sections require fire-resistive construction to be used where exterior walls are located less than 5 feet (1,524 mm) from a property line. The intent of these code provisions is to prevent conflagration or the spread of fire from one building to another by providing "yards" between structures. Additionally, these same code provisions also prohibit openings in walls placed less than 3 feet (914 mm) from a property line. In this case, the latter code restriction creates a conflict with Section R310.1 which mandates an emergency escape opening for bedrooms. Thus, in this unique case, the requirements of the IRC both require *and* prohibit openings in the exterior wall to be located on the property line.

After considerable thought, I suggested the use of a 5 foot (1,524 mm) wide no-build easement as an alternative to the yard required by the code. A no-build easement imposes deed restrictions on a neighboring property that establish a permanent open space in which no construction may take place, creating the equivalent of a yard. Through the use of such no-build easements, the same overall effect can be achieved even though the easement or "yard" is not located on the same property as the proposed dwelling as is required by the IRC. As you have probably surmised, this solution worked because a neighbor was willing, for consideration, to accept the restrictions imposed on her property and because the affected property was largely unusable due to rugged terrain. This technique can be effective, particularly where the terrain or other site characteristics prevent the adjacent property from being developed in the first place, as was the case here.

In this example, no special fire safety systems or construction methods were needed, yet the intent of the code was met through the use of the easement. The no-build easement provided the necessary space adjacent to the dwelling to prevent conflagration, meeting the intent of the code while allowing the proposed dwelling to be constructed right on the property line! Additionally, the width of the easement was deemed to be at least the equivalent to the yard required by the code for unprotected structures. As a result, the emergency openings required by Section R310.1 were allowed, salvaging the project design.

Examples such as this real-world experience illustrate just how far an AMM can deviate from the actual code language yet still provide an effective solution to a code-related problem. Be sure to think creatively and never assume an idea is impractical without first having explored its possibilities. Ask the code official if she has encountered your problem before and what solutions, if any, were found. These discussions may yield solutions to problems in a variety of imaginative ways. After all, a yard by any other name is still a yard!

Other Options

These are but a few of the many options available in a trade-off approach. With a little thought and creativity, it may be possible to find still other solutions in the trade-off approach that will move your green building project forward. Remember, though, that the code official is under no obligation to accept the proposal. She must first find that the trade-off is satisfactory, that it meets the intent of the code, and that it is equivalent to requirements of the code itself. It is essential to involve her in any discussion on a trade-off approach as early as possible to ensure that she knows and understands your intent, is confident in the trade-off approach as an AMM solution, and above all approves its use.

HOW TO MAKE AN AMM REQUEST

Where a product evaluation report is not available or cannot be used, it will be necessary to make an AMM request for alternative forms of construction. As we have seen, AMM requests can be complex and encompass a wide variety of issues. We have also seen that it is essential when dealing with AMM to communicate clearly and completely about the ways in which the AMM meets the intent of the code and meets code equivalency on all levels. I cannot state strongly enough that it is up to the permit applicant to demonstrate this to the code official. Remember, the point of the AMM is to convince someone who, out of necessity, must be cautious and skeptical to accept what are sometimes significant deviations from the code. A thoughtful, organized, well-prepared request with appropriate supporting data is essential in seeking approval for a green building AMM proposal.

So how, exactly, does one make a request to use an AMM? The code official in the jurisdiction in which the permit application is made will almost certainly require a written AMM proposal. The request to use an AMM must not only capture all of the necessary information to address the intent and the equivalency aspects of the code, but will also become a part of an important public record. Where deviations from the basic requirements of the code are granted, she must capture the request and the reasons for the approval in her records.

The local jurisdiction may already have an AMM request form developed and require its use. It would be wise to obtain a copy of it well before making an application for permit. The form will provide clues as to what level of documentation is required for the AMM request and it will also likely identify major areas within the code that must be addressed. These clues will provide at least some idea of what the building official needs to make a decision. Again, the more thorough and complete the application, the more likely one is to receive approval. Remember, the building official must be confident that the AMM proposal meets all aspects of the IRC.

It is also highly advisable to talk with the building official well before making an application. The more involved she is in the AMM process from the beginning, the more she and her team can provide guidance and direction. This is important because it is better to know upfront if there is any chance of getting approval than to waste a lot of time "barking up the wrong tree."

Minimum Information Required for AMM Request

There are, of course, certain basic pieces of information that should be included with any AMM request. Keep in mind that this is a starting place. Your particular set of circumstances may require much more than I have presented here. At the very least, it should contain no less than the following:

Introduction: This can either be made a part of the AMM request itself or provided in a separate cover letter. Here, the goal is to provide an overview of the particular circumstances and a clear reason for making the request. Identify the conditions that create the need to seek approval for the AMM.

Code Requirements: Identify the specific section or sections of the code that apply to the AMM. Include the section and subsection numbers (where applicable) as well as a brief summary of what the code requires. It is also helpful to identify what is believed to be the intent of the section(s), especially if they have already been discussed with the building official.

Specifics of the AMM: Provide detailed information about the specific green materials and/or green construction methods as proposed. Include all relevant information. Remember, the building official may have never seen or heard of the particular product or method being proposed.

Demonstrated Equivalence: This is a key area of the proposal. It must be demonstrated, point by point, how the proposal is equivalent to the code. If the code requires a particular conventional material to meet a certain standard or test, show how the proposed green material does this, or if not, has a trade-off been identified to compensate? If so, clearly state it in the proposal.

Supporting Data: This is also a key area in the proposal. If the product or system being proposed has available test data (recall from Chapter 2 that it should), by all means provide a copy of it in the proposal. If the product manufacturer has obtained a product approval report through the ICC-ES (or has a current legacy report), then provide it. Do not expect the building official to find this information on her own. Include all relevant data; too much information is certainly better than not enough in this case (see Figure 4-6).

Figure 4-6

A construction detail showing the connection of a structural insulated panel at the footing. Note the "sandwich" construction of the panel (plywood/structural foam/plywood). Details such as this are an essential part of an AMM proposal.

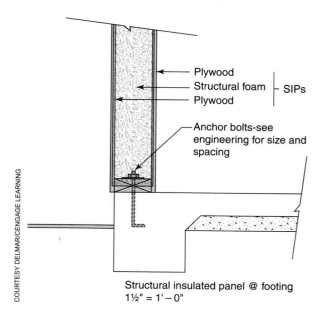

Plywood
Structural foam ⎱ SIPs
Plywood

Anchor bolts-see engineering for size and spacing

Structural insulated panel @ footing
1½" = 1'−0"

COURTESY DELMAR/CENGAGE LEARNING

If you know that the product or system being proposed has been accepted in another jurisdiction, then provide contact information for the building official who approved it. Although acceptance of the AMM by another jurisdiction does not necessarily set a precedent, it may be helpful for all involved to know and understand the basis for that approval. Her experience and familiarity with the AMM may save the building official in your jurisdiction much time and effort.

If the local jurisdiction does not already have an AMM request form, an Alternative Materials and Methods Request Form has been provided in Appendix C of this text, which can be used as a template for any AMM request. This particular form and approach is effective because it requires the person making the request to answer many key questions. These are the same questions typically asked by code officials when processing a request for AMM. Review the form with the building official in the local jurisdiction to ensure that it will meet her needs. It may require slight adaptation to work for one set of circumstances or another.

END OF PART I

We now have a comprehensive overview into the world of building codes. We have broadened our understanding of the basics of code enforcement and learned why communities need effective building codes in the first place. We have also learned the importance of (and how to develop) a strong relationship with the code official in our jurisdiction. Additionally, we learned how the IRC regulates building materials and construction methods and have developed an understanding of just how green building fits into the mix. We are now ready for Part II, which will explore each of the various chapters in the IRC. We will look for opportunities to make green decisions and study how each applicable provision of the IRC applies. Where appropriate, a "green" example will be given using a common green construction method. Now, on to the second part of our journey.

PART II

Part II of this book is intended to provide the reader with a comprehensive analysis of the requirements of the International Residential Code as it relates to green building. It may be used as a stand-alone section or in conjunction with Part I of this text. Review it as often as necessary to become comfortable with the various provisions of the code and to become familiar with the many green decisions to be made in the planning and construction of a green dwelling. Each chapter in Part II of this text corresponds to a specific chapter in the IRC. All references to the "IRC" or "the code" are from the 2009 edition of the International Residential Code, the scope of which covers detached one- and two-family dwellings and townhouses not more than three stories in height.

To help the reader understand the important topics discussed throughout this text, the concepts discussed in each chapter will be applied to a sample project, much like one might encounter in the actual planning of a green construction project. The scope of the sample project is introduced at the end of Chapter 5. The sample project will follow throughout the remaining chapters and will include concepts from Parts I and II of this text.

SITE AND BUILDING PLANNING

This chapter corresponds to IRC Chapter 3, Building Planning

learning objective

To know and understand how the specific provisions of the IRC affect site layout and building planning and to identify areas where green options exist in the code. Additionally, to know and understand how the code impacts the green decision both before and after it is made.

IRC CHAPTER 3—OVERVIEW

If there is one chapter in the IRC that could be described as the "meat and potatoes" of the code, it is Chapter 3, Building Planning. Here, many of the most significant code requirements are established, especially in the area of life safety. Unlike most other chapters in the code that focus on a specific subject area (e.g., foundations, the provisions for which are found in Chapter 4), the building planning and construction provisions address residential code requirements across a broad spectrum of topics.

It is helpful to think of the chapter as divided into two main categories: design criteria and general safety provisions. Approximately one half of the chapter is devoted to the establishment of design criteria, which generally governs the design of the structural aspects of the dwelling. The remainder is devoted to the establishment of safety provisions or code requirements that ensure livability and human comfort. Let's explore each area in detail to aid in our understanding of the code.

Design Criteria

As mentioned previously, the IRC is used throughout the United States, the U.S. Virgin Islands, and in other countries as well. Given its widespread use in coastal areas, mountain regions, seismic zones, climates with extremes of heat and cold, and in every other imaginable environment, it is a wonder that any one document

Figure 5-1

A portion of Figure R301.2(4) Basic Wind Speeds for 50-Year Mean Recurrence Interval.

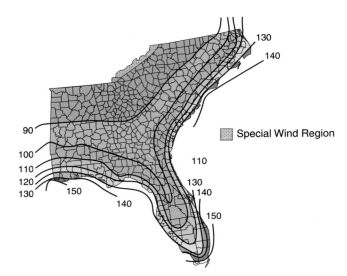

▨ Special Wind Region

For SI: 1 foot = 304.8 mm, 1 mile per hour = 0.447 m/s

a. Values are nominal design 3-second gust wind speeds in miles per hour at 33 feet above ground for Exposure C category.

b. Linear interpolation between wind contours is permitted.

c. Islands and coastal areas outside the last contour shall use the last wind speed contour of the coastal area.

d. Mountainous terrain, gorges, ocean promontories and special wind regions shall be examined for unusual wind conditions.

could provide all of the necessary tools to provide for a safe structure, regardless of the environmental condition.

To its credit, however, the IRC provides flexible criteria so that its provisions can be applied to structures across a broad spectrum of climate types, geographic conditions, and with consideration for regional construction methods. It does so through the use of tables, charts, and maps that establish criteria for ground snow loads, **basic wind speeds** (minimum wind loads used for the design of dwellings, as shown in Figure 5-1), seismic design categories, and even termite infestation probability. These maps and tables are tailored to provide specific criteria for each region of the country, based on years of empirical data and scientific study.

Section R301.1 of the IRC provides a straightforward but profound charging statement as the introduction to the chapter. It provides the scope, if you will, for all structures built under its provisions. It states that "buildings and structures, and all parts thereof shall be constructed to safely support all loads including dead loads, live loads, roof loads, flood loads, snow loads, wind loads and seismic loads prescribed by this code." In plain terms, it means that every structure—green or otherwise—must be designed and built to withstand all structural loading conditions (see Chapters 1 through 4 for a review of dead and live loads if necessary).

Engineered Design

Engineered design is an important consideration for green construction. Because many green construction methods are nonprescriptive, such as straw bale buildings (in those areas where no prescriptive code exists), earth-sheltered structures and others, they must be engineered to meet the design criteria established in the code. This means that someone with special training and credentials must perform structural calculations to determine the adequacy of the proposed construction and whether it is "constructed to safely support all loads" as charged by the code. This typically means that the analysis must be performed by a registered professional engineer or architect licensed to practice in the state where the construction will occur.

Some states require this work to be performed by a licensed *structural* engineer, but others will allow the work to be performed by any engineer who has the requisite skills necessary to do the job. Still others allow the structural analysis for residential structures to be performed by a licensed architect. In any case, the notion of the engineered design and the need to provide engineering should become a familiar part of your vocabulary, as it will almost certainly be a requirement for at least a portion of your green construction project.

Section R301.1.3 of the IRC makes an outright allowance for the provision of engineering. Here, the code permits engineering for all buildings and structures to be provided in lieu of compliance with the conventional, prescriptive construction methods outlined in the code. Where it is provided, the engineered design must be in accordance with the provisions of the 2009 edition of the International Building Code (IBC).

The provisions of this section also allow unusual or otherwise unconventional *elements* (i.e., portions) to be engineered. In other words, it allows a "mix and match" approach to be used; that is, elements of both prescriptive and engineered construction can be combined. In this case, engineering is required only for those portions that are nonprescriptive. The prescriptive elements need only be considered in the engineering analysis to the extent that the nonprescriptive elements (i.e., those that are engineered) impact the design.

General Safety Provisions

The remainder of IRC Chapter 3 establishes requirements that generally pertain to the safety aspects of building design. Here, standards for life safety, fire safety, and livability are provided to ensure that dwellings provide for basic human comfort and that risk of fire is reduced to the maximum extent practical.

The provisions in these code sections establish minimum requirements for important items such as light, ventilation and heating, sanitation, and minimum room sizes. Important safety considerations are addressed as well, as the IRC provides criteria for **means of egress** (exiting), the use of plastics, safety glazing, flame spread, smoke density, and flood-resistant construction. As we will soon see, a number of these provisions can potentially impact the green decisions that must be made during the design and construction of a dwelling. There are also several provisions where green options exist in the code even though they are not expressly stated as such and these questions will be explored as well.

GREEN OPTIONS

Although the IRC does not specifically identify green options per se, a number of code provisions contained within the chapter allow for green decisions to be made by the homeowner, builder, or design professional. These provisions make excellent candidates when considering green alternatives as opposed to conventional construction.

There are also code provisions, however, which can potentially impact or limit the available choices one has for green building. As stated previously, the IRC does not do so to intentionally prevent the use of "green" elements in the construction project. Rather, it does so indirectly by establishing standards that can, at times, be difficult to meet with green materials or methods that are unproven. It is wise to know and understand these because they may impact design considerations at the earliest stages of the project. Still others affect the green decision once it has been made. These code provisions will prescribe a course of action that must be taken once the green choice is made in order to satisfy code requirements.

Code Sections that Allow Green Options

Some code sections allow the homeowner, builder, or design professional to choose from several available options in an effort to comply with the code, some of which could be classified as "green." For example, in areas where termites are prevalent, Section R318 of the IRC identifies code requirements for termite control and provides several options for how this can be accomplished. One such method involves the use of chemical **termiticide** treatments, which use a pesticide specifically for control of termites. If, in the course of planning a green home, the decision has been made to avoid the use of chemicals wherever possible, the code also provides several chemical-free choices (steel framing, for example). It is important to note that they are not identified as green options necessarily, but the choices exist nonetheless.

Code Sections that Limit Green Options

Additionally, a number of code provisions affect the green decision before it is made. That is, the IRC contains code requirements that could potentially limit the green choices one has available or, at the very least, affect the choices one makes in the very earliest stages of the design. These code requirements are not intended to be obstacles or to deliberately prohibit or discourage green building. Rather, they are provided to address important health and life safety concerns which can, at times, conflict with the desires and plans of those who wish to build green. Thus, *all* of the factors affecting the design must be considered early in the planning stage to ensure a balance between green construction and building safety.

To illustrate, let's say that a design professional makes the decision to place the proposed dwelling on the site to take maximum advantage of passive solar energy. This green decision must be made when the project is in its infancy, because placement of the building on the property and the building's orientation to the sun will drastically affect the effectiveness of the passive solar system. Of course, these decisions affect the building design as well, but for now let's focus on the site design. Let's also say that, because of certain site characteristics, it is necessary to place the proposed dwelling very close to one of the property lines to make the backyard as large as possible.

Section R302.1 and Table R302.1 of the IRC establish minimum clearances from property lines and also place restrictions on openings where the exterior walls are less than 5 feet from the property line (see Figure 5-2). As you can see in the table, if the exterior wall of the proposed dwelling is to be located less than 3 feet from the property line, no window or door openings are allowed as prescribed in the code. This restriction could make the passive solar option extremely difficult to accomplish without highly specialized, expensive forms of construction, such as rated windows or door assemblies.

Code Sections that Affect Green Choices Once Made

Finally, the IRC establishes building code requirements that impact the green decision after it is made. In these cases, the code imposes restrictions and standards for construction that will make it necessary to take a particular action to satisfy the

Figure 5-2

Table R302.1 Exterior walls.

TABLE R302.1
EXTERIOR WALLS

EXTERIOR WALL ELEMENT		MINIMUM FIRE-RESISTANCE RATING	MINIMUM FIRE SEPARATION DISTANCE
Walls	(Fire-resistance rated)	1 hour-tested in accordance with ASTM E 119 or UL 263 with exposure from both sides	< 5 feet
	(Not fire-resistance rated)	0 hours	≥ 5 feet
Projections	(Fire-resistance rated)	1 hour on the underside	≥ 2 feet to 5 feet
	(Not fire-resistance rated)	0 hours	5 feet
Openings in walls	Not allowed	N/A	< 3 feet
	25% maximum of wall area	0 hours	3 feet
	Unlimited	0 hours	5 feet
Penetrations	All	Comply with Section R317.3	< 5 feet
		None required	5 feet

For SI: 1 foot = 304.8 mm.
N/A = Not Applicable.

code requirements. For example, one emerging trend in sustainable development and construction is to build smaller dwellings; that is, to construct houses with a smaller "carbon footprint" and less ecological impact overall. In these cases, the IRC contains provisions for minimum room sizes, ceiling heights, and plumbing fixture clearances. In effect, the code imposes restrictions that affect the green decision by establishing just how small the habitable and usable portions of the structure can be.

The remainder of this chapter and future chapters will address some or all of these three scenarios where they exist in an effort to inform the reader of the possible effects of the code's regulations. Eventually, it will become clear how the code impacts the green decisions that must be made and the reader will soon develop a level of familiarity that will enable him to anticipate the impact of the code on the proposed project and, more importantly, to find effective solutions to problems that will inevitably arise during the planning and construction processes.

TAKE NOTE!

Although the focus of this chapter is the building planning section of the IRC (Chapter 3), a number of relevant code provisions from Chapter 1, Administration, impact the planning and design of a green dwelling as well. For the reader's convenience, these administrative provisions are addressed here where they are most relevant and useful.

SITE PLANNING AND RELATED TOPICS

One of the early steps in the design of a green construction project is the planning and decision making associated with the construction site itself. Often, long before the building has even been designed, decisions are made regarding the site and what will be built upon it that impact the entire future of the construction project. It will prove useful in our quest to go green to know and understand how the IRC regulates construction sites—and how it does not regulate them.

For example, there has been an explosion of home-based businesses in recent years. Also known as home occupations or other names, in such cases the dwelling serves as both the place of residence and the business location for the building owner. Home-based businesses are decidedly sustainable because they allow the owner-occupant to live and work at the same location, eliminating or greatly reducing the need to commute to work. Some of the benefits in doing so should be immediately apparent: Less time on the road means decreased vehicle use, less fuel consumed, less pollution, and reduced carbon footprint overall. But the decision to locate, build, and operate a home-based business on the dwelling site can have important design consequences that must be fully considered early on.

Both the 2009 International Residential Code and the 2009 International Building Code now recognize and regulate these types of buildings, referred to as **live/work units** under the provisions of these codes. In addition to these code regulations, local zoning regulations and ordinances will likely come into play as well. Thus, the decision to site and construct a new building that will serve as a live/work unit will have to be carefully considered to ensure code compliance. Is an automatic fire sprinkler system required? Will local ordinance require customer parking? Where will deliveries take place? It should be obvious by now that knowing and understanding the way in which live/work units are regulated under the code and through local ordinance will be important, as these types of structures are designed so that all codes and regulations can be met and so that there will be no surprises later. For a comprehensive discussion of live/work units, see the additional Delmar Cengage Learning title, *Going Green with the IBC,* which will soon be available through www.delmarlearning.com.

Section R101.2 Demolition Permits

Some residential construction projects begin with the demolition of existing structures on the chosen site. Whether these structures are old dwellings, barns, or dilapidated commercial buildings, it is sometimes necessary to remove them or portions of them prior to construction. A number of green options exist that are related to demolition permits and they are explored here.

Section R101.2 of the IRC establishes the need for a demolition permit to be obtained before any structure can be demolished. It is interesting to note that while the IRC establishes the requirements to obtain a permit prior to demolition, it does not provide and has never contained code provisions that govern just how demolition is to be accomplished. Many jurisdictions have specific procedures

Figure 5-3

A large debris pile from a traditional demolition project. Great caution must be used during the demolition process to avoid overhead power lines, gas and electric utilities, and other hazards.

COURTESY OF ISTOCKPHOTO.COM/MIOKOVIC

outlined in their ordinances and policy manuals, however, so it is wise to consult with the code official prior to performing any demolition work.

Why is a permit required, then, if there are no code regulations governing the process? Demolition is, by its nature, a messy process and is potentially dangerous (see Figure 5-3). Since traditional demolition often involves excavation, precautions must be taken prior to the start of any digging. Underground utilities such as electrical services and gas lines must be safely located and terminated or abandoned prior to removal. It stands to reason that cutting through a live electrical service or pressurized gas line can have profound safety consequences.

Structures on adjacent properties may be affected by the demolition process as well, so it is also important to keep the needs of other property owners in mind. Some building materials contain known hazards such as asbestos and must be handled and disposed of in an approved way. Permits ensure that these precautions have been taken into account, even though they may be regulated by an outside agency.

Additionally, it may be necessary to verify that a structure can be legally removed from the site. For example, in areas where there are large numbers of buildings that are landmarks or are otherwise historically significant, it may not be possible to remove them, regardless of condition or plans for future development. It is essential that the reader discuss these matters with the local code official (see Figures 5-4a and 5-4b).

Intent

The intent of this brief but important code provision is to ensure that demolition occurs safely and legally. While the IRC does not prescribe specific requirements for demolition, the code further intends that local jurisdictions have an opportunity to be involved in the process, as other laws and rules may apply.

Green Decisions and Limitations

Shall I demolish the existing structures on the site in the traditional way or deconstruct them in a sustainable way?

How will state laws, local ordinances, and building department policies impact my decision?

Options

Where the requirements of IRC Section R101.2 apply, consider the following green alternatives.

**Figure 5-4(a) &
5-4(b)**

The historical Gordon House,
designed by Frank Lloyd
Wright, now preserved
at the Oregon Garden,
Silverton, Oregon. Often,
state and federal regulations
prohibit the demolition
of historic structures.

(a)

(b)

Recycling: Where recycling is available, it is generally regarded as a more
sustainable way to dispose of an existing structure because it provides an alternative to the landfill. Many construction materials such as concrete, scrap metal,
glass, and wood products can be recycled and made into new materials for use in
the construction or other industries rather than simply being disposed of in the

traditional way. Used roofing materials may be recycled into other products as well.

Recycling centers now exist in or near many major metropolitan areas in the United States. If recycling is not available near your construction site, recycling centers in adjacent communities or states may be a resource. Though it is true that deconstruction can take longer than traditional demolition, careful planning in the early stages of the green project may help overcome this obstacle, particularly where a jurisdiction will not issue a demolition permit until the building permit is issued.

Salvaged materials: In some areas of the country, markets are such that scrap materials can be sold to offset some of the costs related to the deconstruction of the building. Scrap metal in particular can bring a good price in some markets. Lumber can also be salvaged and has a number of uses after it has served a useful life in building. Lumber salvaged from older buildings is often old-growth wood with highly sought after characteristics such as unusually straight grain and high strength. These characteristics make it very desirable for use in millwork, trim, and in other decorative applications, as well as for reuse in certain structural applications (see Figure 5-5). Demand for salvaged lumber can also bring about a premium price in some markets, depending on the application.

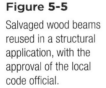

Why It's Green

Deconstruction is green because it allows for many construction materials to be salvaged and either recycled or reused. Hazardous materials such as asbestos-containing flooring and roofing products can be more easily managed and disposed of properly. In addition, considerably smaller amounts of the construction waste end up in landfills as a result.

Figure 5-5

Salvaged wood beams reused in a structural application, with the approval of the local code official.

COURTESY DELMAR/CENGAGE LEARNING

Salvaged materials can often include windows, doors, and unusual and highly decorative millwork such as fireplace surrounds, mantles, and wainscoting. Some of these architectural components cannot be easily replicated today and they make a striking addition to the green home. Salvaging such components can be an easy way to capture the charm and character of eras long past and make a green statement as well, if it can be demonstrated that they will satisfy current code requirements. They are also highly sought after in some markets, as architectural salvage companies exist to acquire and resell these vintage products.

Section R104.9.1 Used Materials and Equipment

Section R104.9 of the IRC provides a charging statement that requires all materials and equipment approved by the building official to be constructed and installed in accordance with such approval. In other words, where the building official has approved a particular building material or piece of equipment for use, it must be installed in the manner in which the building official approved it. Deviations are not allowed, as to do so may change the behavior or performance of the product or piece of equipment—and even affect the dwelling itself—in an adverse way.

Section R104.9.1 specifically addresses *used* materials and equipment, a subject of great interest as it pertains to green building and sustainability. Generally, this section prohibits the reuse of construction materials or equipment unless such use has been approved by the building official. This is important for a variety of reasons. First, used materials or equipment may be inadequate to meet the requirements of today's codes. Second, they may be deteriorated to the point they are unusable or they may be profoundly unsafe in terms of fire resistance. Conversely, the used materials themselves might be the very source of a fire. Third, they may also have unknown physical properties or they might even be dangerous as is the case with asbestos-containing products or other known hazardous materials.

It is important to note that the code does not prohibit the reuse of materials outright. Instead, it simply requires the reuse of materials and equipment to be approved by the code official, which means that it is possible to use them, assuming of course they can be determined to be safe. Still, reusing materials makes sense for a variety of reasons and it is worth pursuing, particularly if the used materials or equipment are in good condition.

remember

Communication with the code official from the earliest stages of planning the green project is the best way to ensure success. Obtain approval before you incorporate used materials into your project.

Intent

The intent of this IRC section is to ensure that materials and equipment are used in the manner in which they were approved. Additionally, the code intends that used materials and equipment be deemed safe for use by the code official prior to their installation.

Green Decisions and Limitations

Shall I reuse materials from my deconstructed building or perhaps from some other site, or shall I use new materials and equipment?

What will the code official require me to do to reuse these materials or pieces of equipment?

Used Lumber Option

Where lumber is allowed to be reused, it must be in generally good, serviceable condition. Where there are obvious areas of decay, say at the end of a beam that has been exposed to the weather, it may be possible to cut off the damaged portion and reuse the shorter component, assuming the remainder is in good shape.

All structural lumber used in construction today must be graded by an approved lumber grading agency. Lumber removed from older structures is not likely to have been graded by the same standards used today, if it was graded at all. At a minimum, the code official will almost certainly require the used lumber to be graded by a qualified individual so that its species and grade can be determined, and thus its structural properties. These properties must then be considered in the overall design of the structure.

Why It's Green

The reuse of building materials, in particular lumber, is a sustainable practice because it allows materials that have already been taken from nature to be given another useful service life. Used lumber, where abundant, can reduce the need for virgin lumber, which can potentially lessen impacts to forests. The amount of lumber products that end up in landfills can be reduced as well.

BUILDING PLANNING

Although the IRC contains provisions that are certainly relevant to the construction site, it is largely concerned with the dwelling. The vast majority of the provisions found in it regulate various components and systems in the building. The safety and comfort provisions provided in Chapter 3 of the IRC will now be explored. Where possible, the impacts of the code provisions on green decisions both before and after they are made will be addressed, and code provisions that afford green options will be explored as well.

Section R302 Fire-Resistant Construction

The exterior walls of all buildings are required to comply with Section R302. This includes not only the construction for the wall itself, but also **projections** beyond the wall (roof overhangs, cantilevers, etc.), **openings** (windows, doors, and vents), and penetrations (pipes, cables, and similar) as well. These elements are required to comply with Table R302.1 which defines when and where openings are allowed, where fire-resistive construction is required, and how far the element must be located from the property line (see Figure 5-2).

This code section and referenced table requires that walls be fire-resistance rated for a period of not less than 1 hour, with fire exposure on both sides of the wall where walls are located less than 5 feet (1,524 mm) from the property line. Fire exposure on both sides of the wall means that a fire is assumed to be burning on either side of the wall (i.e., a fire burning on the interior or a fire burning on the exterior side of a dwelling). Fire-resistance rated construction must meet certain standards for performance. In code parlance, this means that the wall construction must be a **listed assembly** (a wall configuration recognized by an approved standards body) that has been tested to ensure it will meet the 1-hour fire-resistivity requirements.

In certain instances, the underside of projections must also be protected by 1-hour fire-resistance rated construction. Projections may not extend more than 12 inches (305 mm) into areas where openings are prohibited, which is why Table R302.1 requires at least 4 feet (1,219 mm) of fire separation distance from the property line. It is important to note that the requirements for walls and projections do not apply to walls that are perpendicular to the property line.

Openings in the walls, such as windows and doors, are restricted when walls are placed less than 5 feet (1,524 mm) from property lines. When walls are located between 3 feet (914 mm) and 5 feet (1,524 mm) from the property line, the total area of the openings is limited to 25% of the wall area. So, if a wall is located, say, 4 feet (1,219 mm) from the property line and the total wall area is 400 square feet (37.16 m²), the total area of all windows and doors in that wall is limited to 100 square feet (9.29 m²). Where the wall is located at less than 3 feet (914 mm) from the property line, no window, door, or vent openings are allowed. The area of window and door openings is not limited when walls are 5 feet (1,524 mm) or more from the property line.

Penetrations into a fire-resistance rated wall or the underside of a fire-resistance rated projection from pipes, cables, outlet boxes, and similar items are regulated under the provisions of Section R302.4 Dwelling Unit Rated Penetrations. This code section divides penetrations into two categories: **through penetrations** and **membrane penetrations**. As the name implies, through penetrations pass through both the exterior and interior faces of an exterior wall or both sides of a rated interior wall. Membrane penetrations pass through only one face of the wall (see Figure 5-6). Either way, the intent here is to prevent flame and hot gasses generated during a fire from entering the wall. Each category of penetration is carefully regulated under the code.

Intent

The intent of this section is to prevent conflagration, or spread of fire from one building to the next. The IRC intends to reduce this risk by imposing restrictions on how closely walls and projections can be placed with respect to the property line and by requiring fire-resistive construction where these safe limits are

Figure 5-6

An example of a projection beyond the exterior wall (for a gas, direct-vent fireplace) and a through penetration in a nonrated wall.

COURTESY DELMAR/CENGAGE LEARNING

exceeded. Risk is further reduced by the prohibition of openings in walls less than 3 feet (914 mm) from the property line and by restricting the allowable areas of window and door openings where located 3 or more feet (914 mm) but less than 5 feet (1,524 mm) from the property line.

Green Decisions and Limitations

Shall I utilize solar power in some way as a part of my green construction project?

In what way will code restrictions on the placement of walls, projections, openings, and penetrations adjacent to property lines impact my green construction project?

Options

Where solar power is desired and where the requirements of IRC Section R302 allow, consider the following green options.

Passive solar: By carefully planning the location of the proposed building, it may be possible to avoid locating window walls intended to be used for solar energy gain so closely to a property line that the opening areas are restricted or prohibited outright. Where, due to space considerations, it is necessary to locate a wall less than 5 feet (1,524 mm) from a property line thus limiting the availability of windows, it may be

Why It's Green

Incorporating solar power is green because it takes advantage of free energy from the sun and its use produces no harmful waste products that could affect the environment. Solar power reduces the need for the fossil fuels used in traditional heating systems. Some solar panels (photovoltaic systems) actually *generate* electricity.

Figure 5-7

A drawing of a typical, roof-mounted solar water heating system. Water is routed to the solar panels where it is heated and returned to a storage tank.

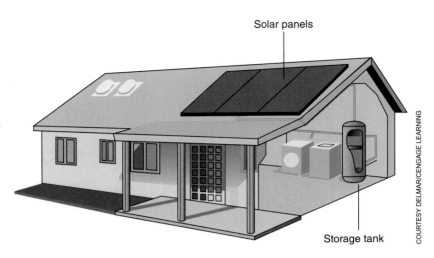

Solar panels

Storage tank

COURTESY DELMAR/CENGAGE LEARNING

possible to orient the dwelling such that the north or east walls are against the property lines. This may allow the south and west exposures typically used for passive solar heating with more distance from the property lines and enable unrestricted openings.

Solar water heating and hydronic space heating: Where passive solar is not an option due to climatic conditions or where the fire-resistive construction requirements noted above are encountered, consider solar water heating and/or hydronic space heating. Solar panels for these systems are typically installed on the roof of a dwelling, making them practical even where openings into walls are prohibited (see Figure 5-7). In some locations where space (and the sun) is abundant, panels may also be located on the ground some distance from the dwelling.

Solar photovoltaics: Another solar option exists through the use of solar photovoltaics. Like passive solar, solar water heating, and hydronic heating systems, photovoltaics utilize free energy from the sun to generate electricity. These systems are typically roof-mount installation, as are other solar panels, and likewise are practical where openings are prohibited in walls (see Figure 5-8). Some photovoltaic systems can generate enough electricity to fully operate a home and can be used to heat water or operate appliances. In some cases, they generate surplus power that, in certain regions, can be sold back to the local electric utility making them an attractive green option.

Section R302.9 Flame Spread and Smoke-Developed Indices

One primary consideration of the code is to limit the speed at which a flame will spread along wall and ceiling surfaces once they are ignited. Additionally, the code is concerned with the development of smoke (in particular, smoke density) when wall and ceiling materials burn. Sections R302.9.1 and R302.9.2 provide criteria that serve as maximum limits for surface burning characteristics and smoke density.

Figure 5-8

An array of photovoltaic solar panels installed on the roof of a dwelling.

COURTESY DELMAR/CENGAGE LEARNING

Section R302.9.1 establishes the maximum allowable flame-spread index at no greater than 200. A vast majority of building materials are tested to determine flame-spread classification using the ASTM E84 tunnel test described in Chapter 2 and are required to meet this important code requirement. Using red oak flooring as a baseline, the rate at which flame spreads across the surface of a material as it burns is compared to the rate at which the red oak specimen burns. The resulting rating is known as a flame-spread classification or index. Red oak flooring is rated with the flame-spread index of 100.

Section R302.9.2 establishes the maximum allowable smoke-developed index for wall and ceiling finishes at no greater than 450. Again, using red oak flooring as a baseline, the density of smoke is measured as the material burns. Smoke densities are measured during the ASTM E84 test using sophisticated equipment. Red oak flooring is rated with the smoke-developed index of 450. As an alternate to the ASTM E84 test, the NFPA 286 test may be accepted under the provisions of Section 302.9.4. The two tests are different, however, and certain other criteria apply. If the green project requires the use of unusual or uncommon wall and ceiling finishes, it is essential to determine the flame-spread and smoke-developed indices. Discuss this with the code official at the first possible opportunity.

Intent

The intent of this IRC section is to ensure that wall and ceiling finishes used in dwellings do not encourage the rapid spread of flames and dense smoke once ignited. The code intends that all wall and ceiling finishes installed in dwellings stay below a certain threshold level. The code further intends that these ratings be determined by widely recognized, acceptable test methods.

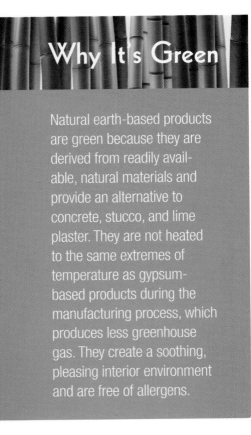

Why It's Green

Natural earth-based products are green because they are derived from readily available, natural materials and provide an alternative to concrete, stucco, and lime plaster. They are not heated to the same extremes of temperature as gypsum-based products during the manufacturing process, which produces less greenhouse gas. They create a soothing, pleasing interior environment and are free of allergens.

Green Decisions and Limitations

Shall I use an alternative interior finish instead of gypsum- or lime-based products for walls and ceilings in my green project?

Will this product meet the IRC requirements for flame-spread classification and smoke-developed index?

Options

When selecting wall and ceiling finishes that will meet the requirements of IRC Section R302.9, consider the following green alternatives.

Earth plaster: Natural earth materials such as earth or clay plasters are gaining popularity. These materials are reminiscent of plasters used centuries ago. Earth and clay plasters create a soothing interior environment and can be finished in a variety of textures, making them extremely versatile (see Figure 5-9). Earth and clay plasters are low VOC, which means they release little to no volatile organic compounds. Also, in some cases, leftover materials can be reused simply by breaking them up and rehydrating them, resulting in less waste. Earth and clay plasters are generally regarded as noncombustible, with a flame-spread classification and smoke-developed index of zero.

Cork finishes: These materials have been around for centuries and come in a variety of colors and textures. Cork is sustainably harvested in a manner that does not harm the tree of origin. Cork produces no off-gasses and does not release fibers into the room environment, thus improving indoor air quality—an important consideration in the green dwelling. Cork has a flame-spread index of 25 and a smoke-developed index of 250, per ASTM E84 tests. This performance level is well below the FS200/SD450 required by the IRC, making it suitable for use in dwellings.

Figure 5-9

Clay earth plaster used as a wall covering in a bathroom.

COURTESY OF AMERICAN CLAY

Section R303.8 Required Heating

This IRC section requires all dwellings to be provided with heating facilities where the winter design temperature is below 60° F (16° C). Figure R301.2(1) in the design criteria portion of IRC Chapter 3 identifies all areas within the continental United States as having a design temperature below that threshold, thus a heating system is generally required in all dwellings (see Figure 5-10).

Figure 5-10

Figure R301.2(1) isolines of the 97.5% percent winter (December, January, and February) design temperatures (°F).

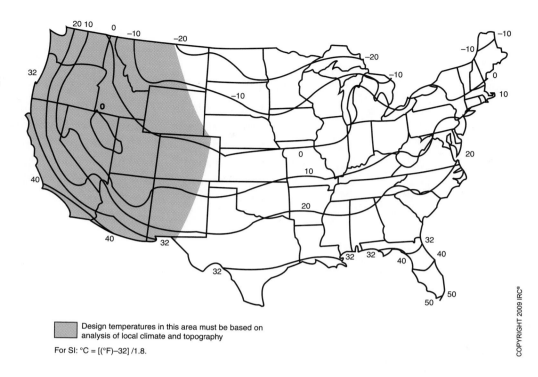

Design temperatures in this area must be based on analysis of local climate and topography

For SI: °C = [(°F)−32] /1.8.

In many of the western states, however, the code requires the winter design temperatures to be determined by analysis of local climate and topography. Where it can be demonstrated to the code official that the winter design temperature is above 60° F (16° C), it may be possible to omit the required heating system. Be sure to discuss this AMM with the code official where plans for the green project do not include a traditional heating system such as a forced-air furnace or heat pump as might be the case with a **net-zero dwelling** (a dwelling that generates at least as much energy as it uses). It will be necessary to support the request for AMM consideration with appropriate scientific data. Where the winter design temperature is below 60° F (16° C) or where it cannot be sufficiently demonstrated that the winter design temperature is above that threshold, be prepared to provide a mechanical heating system.

remember

The burden of proof lies on the permit applicant when preparing an AMM proposal, so be sure to do your homework!

Required heating systems must maintain a room temperature of 68° F (20° C) at the required outdoor design temperature. In habitable rooms, this room temperature is determined 3 feet (914 mm) above the floor and 2 feet (610 mm) from exterior

walls. The IRC is clear that portable space heaters may not be used to achieve compliance with the requirements of this code section.

Intent

The intent of this IRC section is to maintain a safe, comfortable indoor temperature in areas where winter temperatures are deemed to be cold environments. Additionally, this section intends to protect building systems such as piping from damage due to extremes of cold by conditioning the interior space to a minimum threshold level. This section further intends that required heating be accomplished with permanent rather than portable heating equipment.

Green Decisions and Limitations

How shall I provide heat for my green home?

Does the IRC impose code restrictions that prohibit the use of certain heating systems or require the use of others?

Options

When selecting a heating system to satisfy the requirements of IRC Section R303.8, consider the following green options.

Geothermal heating: Also known as ground-source heat pumps, geothermal heating systems utilize the relatively stable temperatures from the ground inside the earth as the source for heating and cooling. Although temperatures at the surface of the ground and even several feet below the surface can vary widely based on geographic and climatic conditions, temperatures tend to be relatively constant below certain depths. These relatively constant temperatures allow geothermal systems to operate more efficiently because they are not subject to extremes of temperature as are conventional heat pumps. Geothermal heat pumps simply extract heat from the earth and transfer it to the structure using electricity instead of by burning a fuel (see Figure 5-11).

Hydronic heating: Sometimes called radiant floor heat, hydronic heating systems utilize a series of tubes embedded in standard, lightweight, or gypsum concrete to move heated water throughout the dwelling. Heat, then, radiates from the floor as opposed to traditional heating systems which moves heated air through ducts. Some hydronic systems utilize traditional radiators for heating as well. Hydronic heating systems can be combined with solar water heating systems for increases in efficiency. Hydronic systems can improve the quality of indoor air because they do not blow dust and other contaminants around inside the structure; however, fresh air ventilation systems may also be necessary to ensure good indoor air quality. These systems can either be combined with the potable water supply or operate as independent, stand-alone systems.

Section R304 Minimum Room Areas

The IRC establishes minimum sizes for habitable rooms in dwellings, and every dwelling is required to comply with these provisions. Habitable rooms are defined as spaces used for living, sleeping, eating, and cooking. They typically do not

Figure 5-11

A drawing depicting a typical geothermal or "ground source" heat pump. Water is routed through piping into the ground where it is either heated or cooled and returned to the dwelling.

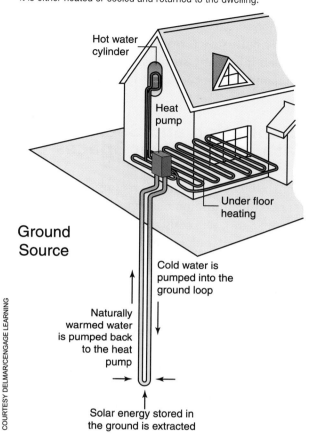

Hot water cylinder

Heat pump

Under floor heating

Ground Source

Cold water is pumped into the ground loop

Naturally warmed water is pumped back to the heat pump

Solar energy stored in the ground is extracted

COURTESY DELMAR/CENGAGE LEARNING

include bathrooms and toilet rooms, halls, closets, laundry rooms, or other utility spaces.

Every dwelling is required to have one habitable room with not less than 120 square feet (11.15 m²) of gross floor area. With the exception of kitchens, other habitable rooms such as bedrooms or dining rooms must provide at least 70 square feet (6.5 m²) in floor area. Additionally, habitable rooms (again, other than kitchens) must have a minimum horizontal dimension in any direction of 7 feet (2,134 mm). Thus, the minimum dimensions for a bedroom to satisfy both the minimum room dimension and room area requirements as required by the IRC are 7 feet (2,134 mm) by 10 feet (3,048 mm). Other room sizes are obviously possible as long as both criteria are met. It is important to note that kitchens do not have a minimum dimension or floor area requirements under the provision of the code.

In habitable rooms that contain sloping ceilings or where the ceiling is furred down to accommodate ducts, beams, or other obstructions, the IRC imposes restrictions on areas where certain minimum ceiling heights cannot be met. In the case of sloping ceilings where the ceiling slopes to less than 5 feet (1,524 mm) above the floor, the area with less than 5 feet (1,524 mm) height clearance cannot be counted toward meeting the minimum room area or room dimension requirements (see Figure 5-12). Similarly, where the height beneath a furred ceiling is less than 7 feet (2,134 mm), it also cannot be included in the room dimension requirements.

Intent

The intent of this IRC section is to provide a minimum level of living space in each dwelling and in habitable rooms for basic human comfort. Additionally, the code further intends that portions of rooms with low ceiling heights not be included in the minimum room area and dimension requirements because such spaces have limited usability.

Green Decisions and Limitations

Shall the size of my dwelling be limited to reduce its "carbon footprint" or impact on the environment?

Does the code impose restrictions that will impact the size of the proposed green dwelling?

Reduced Building Footprint Option

Dwellings that are smaller in size use less materials during the construction process and can often be operated more efficiently than their larger counterparts. It stands to reason that the smaller a dwelling is the less material is needed to build

it. Additionally, smaller dwellings can be more efficient to operate because the smaller volume of space requires less energy to condition it. Long-term maintenance costs can also be reduced with careful planning. After all, less paint required for exterior maintenance and less roofing to replace cannot be a bad thing! This can be an advantage even for conventionally built structures because it is based on a simple but effective idea—an inherently *sustainable* idea—to simply use less of what exists in nature in the course of development.

Section R306 Sanitation and Section R307 Toilet, Bath, and Shower Spaces

These IRC sections establish the basic sanitation requirements for all dwellings and also prescribe the minimum clearances required at all plumbing fixtures (see Figure 5-13). The code requires every dwelling to be provided with at least one water closet (toilet), a lavatory (bathroom sink), and either a bathtub or a shower. Additionally the IRC requires every dwelling to contain a kitchen area and each kitchen area to be provided with a sink. Note that there is no requirement for a laundry facility to be provided in a dwelling.

Plumbing fixtures in dwellings are required to be connected to a sanitary sewer system or to a private sewage disposal system. This typically means that the dwelling must be connected to a municipal sewer or to an onsite waste disposal system such as a septic tank and drain field.

Similarly, every dwelling is required to be connected to an approved water supply. This is typically accomplished by connection to a municipal water supply, an onsite well, community well, or other approved system. A hot and cold water supply is required for all kitchen sinks, bathroom sinks, bathtubs and showers, and where provided, laundry facilities.

Intent

The intent of these IRC sections is to ensure that every dwelling is provided with a certain basic level of sanitation and that there are appropriate clearances around plumbing fixtures to ensure that they are usable and can be maintained. Further, the code intends that all plumbing fixtures be connected to an approved sewage system to ensure that human waste is properly disposed

Figure 5-12

Sloping framing members in a bedroom. That portion of the room under the sloped framing with less than 5 feet (1,524 mm) head room clearance would not be included in the minimum room dimensions.

COURTESY DELMAR/CENGAGE LEARNING

Figure 5-13

Figure R307.1 Minimum fixture clearances.

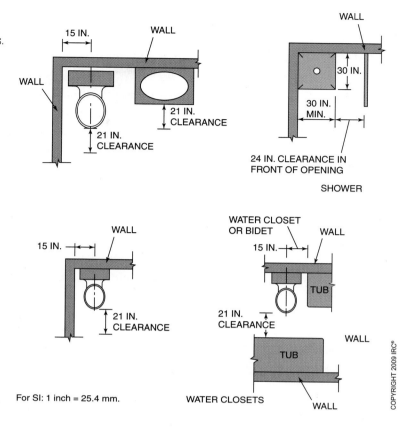

For SI: 1 inch = 25.4 mm.

of for health and safety reasons. Water supplies must be from an approved source and for most plumbing fixtures a supply of both hot and cold water is required.

Green Decisions and Limitations

Can I include alternative plumbing fixtures in my green dwelling?

Do the IRC-imposed minimum sanitation requirements impact my choices for green plumbing?

Options

When selecting plumbing systems to satisfy the requirements of IRC Sections R306 and R307, consider the following green alternatives.

Low-flow fixtures: Since 1992, with the passage of the Federal Energy Policy and Conservation Act, low-flow plumbing fixtures have been required in all new construction. These fixtures save billions of gallons of water each year. Low-flow toilets, showerheads, and faucets, combined with newer, more efficient clothes washing machines, enable Americans to save precious water resources (a decidedly green thing to do) and reduce the embodied energy necessary to produce the water in the first place. In recent years, even more efficient plumbing fixtures

Why It's Green

Alternative plumbing fixtures and systems save water, use less energy, and in some cases reduce the amount of plumbing waste discharged from the building. Some systems harvest water that would otherwise go to waste for use in irrigation systems and for flushing toilets.

have emerged and are gaining popularity. Many of these products will satisfy the basic sanitation requirements of the IRC. These and other options will be explored more fully in Chapter 15.

Rainwater harvesting: A relatively new and exciting trend has emerged in recent years that allows rainwater to be harvested and used for irrigation. These systems are already approved in several states. Where used, these systems can harvest and store rainwater for use during dry seasons, thereby reducing demands for potable water used as irrigation. These and other related systems will also be explored more fully in Chapter 15.

Section R310 Emergency Escape and Rescue Openings

There is probably no more fundamental life safety requirement in IRC than that of the emergency escape and rescue opening. IRC Section R310.1 establishes the requirement for at least one operable emergency escape and rescue opening (also called an egress window) to be provided in every sleeping room and in basements.

These openings serve two purposes. First and most important, the egress window serves as a means of escape directly from the bedroom. This is important because, while the occupants of a dwelling are asleep or otherwise distracted, there is a risk that a fire could grow unchecked, making it impossible to exit through the main portion of the dwelling to access the exit door. In basements, where the walls are below grade, there is a possibility that the occupants of a dwelling could become trapped if the stairs from the basement become blocked by smoke and fire. Fortunately, working smoke alarms to provide early detection and warning in combination with the requirement for the emergency escape openings have proven to be highly effective at reducing fire deaths over time.

There is, however, another important but sometimes overlooked purpose for the emergency escape opening. This opening also serves as a means of access for firefighting personnel. The minimum dimensions and size of the rescue opening are established so that a firefighter wearing protective gear and a breathing apparatus can enter the house to conduct rescue operations (see Figure 5-14). Were it not for these minimum dimensions, access to a dwelling might be impossible under certain fire conditions.

Figure 5-14

A firefighter entering a burning dwelling through an emergency escape and rescue opening. Note the breathing apparatus on the firefighter's back.

COURTESY DELMAR/CENGAGE LEARNING

Emergency escape and rescue openings are required to meet certain minimum opening sizes. Openings above the grade floor are required to open a minimum of 5.7 square feet (0.53 m²) in area. Where these openings occur at or below the grade floor, the minimum opening area is reduced to 5 square feet (0.46 m²). The minimum net opening height required is 24 inches (610 mm). The minimum net opening width required is 20 inches (508 mm). It might be apparent that these two dimensions, when multiplied together, do not produce an opening that will meet the minimum area requirement. Thus, where one dimension is reduced to the minimum, it is necessary to adjust the other sufficiently so as to produce an opening that will comply with the minimum area requirements for the section.

Intent

The intent of this IRC section is to provide a means of emergency escape from sleeping rooms and basements for use during an emergency. The code further intends that each emergency escape opening meets minimum criteria for opening area and opening dimensions. The IRC also intends that each basement and sleeping room be provided with an opening to serve as a means of access for firefighting personnel to conduct rescue operations.

Green Decisions and Limitations

Shall I choose an earth-friendly construction method for my green home, such as earth-sheltered construction?

How will the IRC requirements for emergency escape and rescue openings impact my green project?

Earth-Sheltered Dwelling Option

Certain types of structures, by their nature, make meeting the requirements for emergency escape and rescue openings more difficult than others. In cases where significant portions of the dwelling are below grade, these IRC requirements can pose interesting challenges. Earth-sheltered structures (also called earth berm dwellings) typically are constructed to have at least one side exposed to daylight (see Figure 5-15). Space permitting, such buildings can be laid out in the planning stages to be long and shallow rather than wide and deep, which allows rooms to be pulled forward toward the daylight walls, enabling a window or door to be provided.

Where space limitations prohibit this approach, window wells can be provided at the required emergency escape and rescue openings. Section R310.2 provides the criteria for window wells. Where provided, window wells

Why It's Green

Earth-sheltered construction is green because it uses the relatively constant temperatures below grade to protect the dwelling from extremes of temperature. Earth-sheltered dwellings are insulated by layers of earth, which make them highly energy efficient. Additionally, earth-sheltered structures minimize the impact on the landscape because significant portions of the dwelling are hidden "below grade." Landscaping occurs not only around the structure but also atop it.

Figure 5-15

An earth-sheltered structure. Note how the building nestles seamlessly into the earth berm. In this case, the choice of building materials causes it to blend perfectly with its surroundings.

must provide a minimum horizontal area of 9 square feet (0.84 m²) and must be at least 3 feet (914 mm) by 3 feet (914 mm) in dimension. The window well cannot obstruct the opening of the window and must allow it to be fully opened. Where the window well is deeper than 44 inches (1,118 mm), it must be provided with a permanent ladder or steps to enable the room occupants to climb out in an emergency. Required ladder dimensions are 12 inches (305 mm) in width, with rungs spaced more than 18 inches (457 mm) apart. Ladders must project at least 3 inches (76 mm) from the wall on which they are mounted.

Section R317 Protection of Wood and Wood-Based Products against Decay

When wood is exposed to the weather, subject to insect damage, or in cases where experience has shown that environmental conditions will cause decay, the IRC imposes requirements to ensure that it will be adequately protected from damage. IRC Section R317.1 identifies specific conditions and locations in which wood must be protected through the use of chemical preservatives or through the use of wood products that have been determined to be naturally resistant to decay.

Some of these locations and conditions include where wood framing is supported on masonry or concrete foundations less than 8 inches (203 mm) above the ground, where wood sills and sleepers rest on concrete slabs, and at the ends of wood girders entering masonry or concrete walls without sufficient clearances. Wood that is continuously exposed to the weather and supporting decks and roofs must be treated as well in some cases. It is widely known that wood will decay over time when exposed to moisture repeatedly or for prolonged periods. Thus, the IRC makes an attempt to identify those conditions under which this is most likely to occur.

Wood products are especially vulnerable when they are in contact with the ground or embedded in concrete that is in contact with the ground. In these cases, the code requires that wood products be treated with a preservative and be suitable for ground contact use. In these cases, wood is typically **incised** (scored) and preservatives are injected into the wood fiber under pressure to ensure maximum penetration of the preservative.

As is the case with many of the IRC requirements, Section 317.1.3 allows for adjustments in these code provisions based on geographic conditions and where experience has demonstrated a need. It stands to reason that there are certain areas of the United States where moisture is not an issue (desert climates) and where it is a significant issue (coastal areas). These provisions give the local code official the flexibility needed to establish requirements based on local conditions. Discuss these requirements with the local building official.

Intent

The intent of this IRC section is to ensure that structural wood products are adequately protected from decay. Additionally, the code intends that local geographic and climatic conditions be taken into account when establishing requirements for protection against decay.

Green Decisions and Limitations

Can I avoid the use of chemical preservatives in my green dwelling?

How do the IRC requirements impact my decision to avoid chemicals?

Options

Where the requirements of IRC Section R317 apply, consider the following green options.

Naturally durable wood: Certain species and types of wood products have been determined to be naturally resistant to decay. That is, they are resistant to decay without the use of chemical preservatives, which makes them a great choice for a green building project. The IRC defines naturally durable wood as the heartwood of decay-resistant redwood, cedars, black locust, and black walnut. The use of heartwood is an important distinction here as sapwood for most species is less durable than its heartwood counterpart. The code does make an allowance for corner sapwood, however. Where corner sapwood does not comprise more than 10% of a particular side of a piece of lumber (90% heartwood), it is acceptable under the code.

One important point needs to be made here: The IRC allows naturally durable wood to be used only in cases where there is no ground contact. As noted above, ground contact conditions are subject to more severe effects from soil moisture and microorganisms; thus, naturally durable wood products may not be allowed in all cases. This is a great opportunity to engage in a dialogue with the building official about your preferences and goals for the project.

Why It's Green

Chemical-free methods of protection against decay are green because they avoid the use of potentially toxic chemicals which might enter the dwelling or seep into the ground.

Other chemical preservatives: If it is the case that a chemical preservative is deemed mandatory, then there are a number of chemical preservatives in use today. Some of them are more "environmentally friendly" than others and it may be the case that one of these preservative types will be suitable for your green project. Borates, for example, are effective treatments for wood products. Borates are safe and can be handled without the concerns associated with other preservative chemicals. Borates cannot be used in wet conditions or where they will be subjected to ground water, as the chemical preservative will eventually leach out of the product.

Section R318 Protection against Subterranean Termites

Ahhh . . . the lowly termite. As small as that creature is, it can cause significant damage to wood frame structures. The IRC establishes code provisions for the protection of structures from termite damage. As with other requirements, the IRC recognizes that not all areas within the United States are subject to damage from these pests; thus it establishes requirements for protection based on geographic conditions as seen in Figure 5-16.

In addition to wood damage, termites burrow through foam plastics in an effort to find food. In areas of the country where the probability of termite damage is "very heavy," the IRC prohibits use of foam plastics on the outside of the structure and under foundations below grade.

A number of options are provided for the control of termites. Chemical termiticide treatments are allowed and can include soil treatments and/or field-applied

Figure 5-16

Figure R301.2(6) Termite infestation probability map.

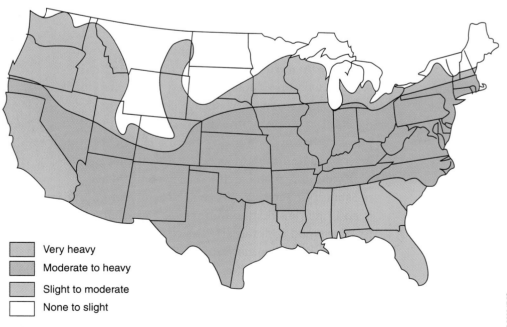

Very heavy
Moderate to heavy
Slight to moderate
None to slight

NOTE: Lines defining areas are approximate only. Local conditions may be more or less severe than indicated by the region classification.

wood treatments. Termiticides are required to be applied in accordance with the recommendations on the product label. Termite baiting systems are also allowed when they are installed and maintained according to the product label.

As with the provisions of Section R317, Protection of Wood and Wood-Based Products against Decay, pressure-preservative-treated wood products are also allowed as a means of controlling termites. Naturally termite-resistant wood products are allowed as well; however, heartwood of redwood and eastern red cedar are the only two lumber products that are recognized by the code as being naturally resistant to termites. Physical barriers are also allowed, but per the provisions of Section R318.3, barriers may be used only where they are combined with another method.

Intent

The intent of this IRC section is to minimize the risk of damage to structures from termites. The code further intends that a means of protection be provided consistent with one of the five methods outlined in this section.

Green Decisions and Limitations

Shall I use a chemical-free option for the control of termites in my green dwelling?

How will IRC requirements affect that decision?

Options

Where the requirements of IRC Section R318 apply, consider the following green alternatives.

Naturally resistant wood: Heartwood of redwood and eastern red cedar lumber products are suitable solutions under the provisions of the code. These products are chemical-free options, which make them a great choice for a green project. Bear in mind, however, that these types of lumber products may not be readily available in all areas of the country. Also keep in mind that each species of wood and the grade of the material determine its structural properties. If these wood products have significantly lower structural properties than wood products typically used to build in your area, structural components such as studs, joists, beams, and girders may need to be upsized as a result, increasing costs.

Physical barriers: Another green option exists in this method of termite control. Recall that barrier systems are required by the IRC to be combined with another method of control. Thus, physical barriers could be combined with baiting systems or naturally termite-resistant wood as a chemical-free method of termite control.

Why It's Green

Chemical-free methods of protection from termite damage are green because they avoid the use of potentially toxic chemicals that might affect the health of the building occupants or seep into the ground, adversely impacting ground water. Additionally, some wood-free methods of controlling termites reduce the need for lumber products, which can help to sustain forests.

Metal framing: Yet another option exists with metal (steel stud) framing. Because organic wood products are the primary food source for termites, constructing the dwelling out of metal framing virtually eliminates their source of food. Metal framing has the added advantage of reducing the need for wood products, which is good for forests and is fully and infinitely recyclable. Provisions for prescriptive metal framing are provided in Chapters 5, 6, and 8 of the IRC. Metal framing provides yet another solution to termite control that should be discussed with the local code official.

SAMPLE PROJECT—INTRODUCTION

The following sample project will be used to demonstrate the ideas presented in the text. The project will be kept fairly simple so that the reader can focus on the application of the code rather than keeping track of complexities that serve no useful purpose. In this case, the project will consist of a simple, single-story dwelling with exterior walls constructed of rammed earth (an AMM) and will utilize light-gauge steel framing where possible. The project will be built on a relatively flat, empty lot. For this exercise, however, let us assume that an unusual site characteristic will require one exterior wall of the sample dwelling to be constructed 3 feet (914 mm) from a property line.

Design Philosophy

As we have alluded to in the preceding chapters, there are many and varied opinions about what makes a project green. There is no question that the "greenness" of the identified items could be debated depending on one's beliefs, philosophical perspective, and goals. Each of the selections made for this exercise could just as easily be replaced with another equally green choice. The purpose of this exercise, however, is to learn how to apply the material covered in each chapter in an effective way. To that end, I ask that you accept the identified scope of work and design philosophies as "green" for illustration purposes. See the Afterword section at the end of the book for more discussion on this matter.

For this sample project, assume the following design philosophies.

- The project should be planned in such a way that it will have minimal site disruption and in a manner that requires as little grading as possible.
- The project will include a concrete slab foundation.
- The project should use as little wood as possible.
- Where it is absolutely necessary to use wood-based products, the sample project will utilize engineered lumber, in particular those products that utilize wood scrap or waste material in their manufacture.
- The dwelling is to be as energy efficient as is reasonably possible.
- The sample dwelling will use local materials when possible, to not only support the local economy but also reduce the environmental impacts of transporting building materials over long distances

- The project will use recyclable materials wherever possible.
- The project should be designed such that the building can be deconstructed rather than demolished at the end of its useful life.
- The project should incorporate prescriptive codes to the maximum degree possible, but also recognize that AMM and engineering will be necessary for the rammed-earth walls.

Scope of Work

The following list identifies the scope of work for the sample project, which will be a one-story, rammed-earth structure. Each aspect of the construction will be addressed in the appropriate chapter.

- For simplicity, the sample construction site is assumed to be generally flat and will require little grading or preparatory work.
- The exterior walls will be constructed of stabilized, reinforced rammed earth.
- Roof and interior wall framing will be constructed of light-gauge steel framing.
- Insulation for the roof and other areas requiring it will be of the spray-applied foam type.
- The sample dwelling will utilize the unvented attic option outlined in IRC Section R806.4.
- Windows and doors will be selected for maximum energy efficiency.
- The mechanical system will incorporate a high-efficiency, geothermal heat pump.
- The plumbing system will incorporate ultra low-flow fixtures and a central, tankless water heating system.
- The electrical system will incorporate high-efficiency lighting and solar photovoltaics.

CHAPTER 5 APPLIED TO THE SAMPLE PROJECT

Chapter 5 addressed matters related to the planning of the site and important decisions to be made early in the design process. Additionally, Chapter 5 addressed many life safety code requirements. The following analysis will demonstrate the application of some of the IRC provisions with consideration given for the proposed scope of work and our design philosophy.

Site Considerations

The projection description tells us that the site is relatively flat and empty. The empty site means that no demolition permit will be required initially; but also notice that in our design philosophy we want to construct a dwelling that can be easily deconstructed later. The metal framing (with screw-type fasteners) and the use of rammed earth will make this easier to accomplish in the future. Given that we want to minimize grading and site work, the flat lot is well suited to the task. To be safe, we will check with the local jurisdiction to see if a grading permit is required and take the necessary steps to satisfy any requirements.

We are also told that it will be necessary to locate one exterior wall of the dwelling 3 feet (914 mm) from a property line, so we know from Section R302.1 and Table R302.1 of the IRC that 1-hour rated fire-resistive construction will be required for this exterior wall. We also know from the same table that unprotected openings will be allowed but are limited to 25% of the wall area. This will enable us to locate bedrooms along the rated wall in our design and to provide the emergency exit windows required by Section R310.1, provided the total area of all openings does not exceed 25%.

Building Considerations

The project description tells us that the exterior walls of the dwelling are proposed to be stabilized (cement added), reinforced rammed earth, the soil for which we will obtain locally if possible. We know after careful review of the IRC that the code contains no prescriptive provisions for rammed-earth construction, so clearly this will fall into the category of an AMM. We will contact the code official immediately to discuss the AMM and to find out from him what information must be submitted if the project is to be successful.

At the very least, the AMM will need to demonstrate that the rammed earth meets the intent of the code and is equivalent in every aspect. From this, we know that we must demonstrate through test and other data that the rammed-earth wall will meet at least the 1-hour fire-resistive requirements imposed by the code. We also know that the rammed-earth wall must be able to resist all structural loading requirements and be fully engineered. Our engineering and test data, then, must be thorough and comprehensive to ensure that we can satisfy all code requirements. Careful and thorough research will be required so that the design professional will have adequate specifications and criteria from which to perform the structural analysis.

CHAPTER SUMMARY

Chapter 3 of the IRC covers a broad array of topics, many of which are critical to know early in the design process. Additionally, there are a number of code provisions that come from Chapter 1 of the IRC that impact the design, particularly in the layout of the construction site. Following are highlights of the chapter.

- Chapter 3 provides design criteria through maps, tables, and figures that are used to design the structure of a dwelling. It also identifies regional information relevant to the design of a dwelling.
- Chapter 3 provides important life safety, fire safety, and human comfort criteria that must be considered in the design of a green dwelling.
- Some code provisions afford choices or options for code compliance, some of which can be made in a green way.
- Other code provisions impact or restrict the choices that must be made early in the design process but not intentionally. Instead, the code establishes requirements that sometimes conflict with and take precedence over green considerations.
- Still other code provisions require the homeowner, builder, or design professional to do certain things once the green decision is made.
- Some code provisions impact the site layout and must be made during the initial planning stages of a construction project.
- The majority of code IRC provisions are specific to the dwelling (i.e., the building) itself, as opposed to the building site.

FOUNDATIONS

This chapter corresponds to IRC Chapter 4, Foundations

learning objective

To know and understand how foundations are regulated under the International Residential Code (IRC). Also, to explore green options that exist in the layout and planning of a site and in the design of a foundation.

IRC CHAPTER 4—OVERVIEW

All structures, regardless of by what method they are built, must be provided with a suitable foundation. The foundation is the interface between the building and the earth on which it rests. All loads that develop within the structure, from gravity loads to lateral forces, are ultimately transferred to the foundation. Thus, it is vitally important that the foundation be designed and constructed in a way that allows it to resist these forces and enable the dwelling to remain standing when subjected to a variety of loading conditions.

Chapter 4 of the IRC provides the design criteria for foundations and specifies both materials standards and prescriptive methods of construction for a variety of foundation types to ensure that foundations constructed under its provisions will safely support the structure superimposed on top of it. These criteria are provided in the body of the chapter (i.e., the text) or in tables like the one shown in Figure 6-1.

As we have seen in other chapters, Chapter 4 regulates both materials and construction methods. Construction materials for several different types of foundations are addressed here. For example, the IRC provides criteria for the use of wood foundations which, although used infrequently, provide a viable option where it is impossible to use concrete or masonry products. Wood foundations resemble ordinary wood frame walls except that a portion of the wall occurs below grade (see Figure 6-2).

The IRC also provides criteria for concrete footings and foundations, which are used extensively in residential construction. Concrete is an excellent material for use in foundations, as it is extremely strong and durable. Its structural properties, like compressive strength, can be carefully controlled by regulating the proportions

Figure 6-1

Table R404.2.3 Plywood grade and thickness for wood foundation construction.

TABLE R404.2.3
PLYWOOD GRADE AND THICKNESS FOR WOOD FOUNDATION CONSTRUCTION
(30 pcf equivalent-fluid weight soil pressure)

HEIGHT OF FILL (inches)	STUD SPACING (inches)	FACE GRAIN ACROSS STUDS			FACE GRAIN PARALLEL TO STUDS		
		Grade[a]	Minimum thickness (inches)	Span rating	Grade[a]	Minimum thickness (inches)[b,c]	Span rating
24	12	B	$15/32$	32/16	A	$15/32$	32/16
					B	$15/32^c$	32/16
	16	B	$15/32$	32/16	A	$15/32^c$	32/16
					B	$19/32^c$ (4, 5 ply)	40/20
36	12	B	$15/32$	32/16	A	$15/32$	32/16
					B	$15/32^c$ (4, 5 ply)	32/16
	16	B	$15/32^c$	32/16	B	$19/32$ (4, 5 ply)	40/20
					A	$19/32$	40/20
					B	$23/32$	48/24
48	12	B	$15/32$	32/16	A	$15/32^c$	32/16
					B	$19/32^c$ (4, 5 ply)	40/20
	16	B	$19/32$	40/20	A	$19/32^c$	40/20
					A	$23/32$	48/24

For SI: 1 inch = 25.4 mm, 1 foot = 304.8 mm, 1 pound per cubic foot = 0.1572 kN/m³.

a. Plywood shall be of the following minimum grades in accordance with DOC PS 1 or DOC PS 2:
  1. DOC PS 1 Plywood grades marked:
    1.1. Structural IC-D (Exposure 1)
    1.2. C-D (Exposure 1)
  2. DOC PS 2 Plywood grades marked:
    2.1. Structural I Sheathing (Exposure 1)
    2.2. Sheathing (Exposure 1)
  3. Where a major portion of the wall is exposed above ground and a better appearance is desired, the following plywood grades marked exterior are suitable:
    3.1. Structural I A-C, Structural I B-C or Structural I C-C (Plugged) in accordance with DOC PS 1
    3.2. A-C Group 1, B-C Group 1, C-C (Plugged) Group 1 or MDO Group 1 in accordance with DOC PS 1
    3.3. Single Floor in accordance with DOC PS 1 or DOC PS 2

b. Minimum thickness $15/32$ inch, except crawl space sheathing may be $3/8$ inch for face grain across studs 16 inches on center and maximum 2-foot depth of unequal fill.

c. For this fill height, thickness and grade combination, panels that are continuous over less than three spans (across less than three stud spacings) require blocking 16 inches above the bottom plate. Offset adjacent blocks and fasten through studs with two 16d corrosion-resistant nails at each end.

Figure 6-2

An illustration showing a wood foundation basement wall. Note the use of foundation grade lumber and sheathing below grade.

Joist hangers

Angle brackets

2 x 8 Foundation grade treated wood studs at 16" on center

19 x 32" Foundation grade plywood

12" Treated backfill board

Waterproof membrane

Granular backfill drain rock

2 x 10 Treated footing plate

Compacted gravel footing

Sump pump

Sump basket

Footing width is 16"

Sheathing fastened with 8d stainless steel nails

Framing fastened with 16d galvanized nails

COURTESY DELMAR/CENGAGE LEARNING

of the raw materials used when the concrete is manufactured. This control makes the behavior of concrete highly predictable, making it an ideal material for use in both prescriptive and engineered foundation configurations. Concrete footings and foundations can be reinforced with **reinforcing steel** (rebar) which provides additional strength, particularly when concrete is loaded in tension as is the case with certain footings, beams, and retaining walls. In some geographic areas where seismic activity is not a concern, plain concrete foundations are allowed and provisions are made in the IRC for these conditions well.

Like concrete, masonry may also be used for footings and foundations. **Concrete masonry units** (concrete block or CMUs) are manufactured in modular, blocklike units. These block units are stacked in various patterns to form footings and walls. Masonry blocks are typically made of concrete and, when properly reinforced and grouted, are comparable to poured concrete in terms of strength and stability.

Unlike concrete, however, masonry units are preformed units that are modular in nature and do not require forms as they are placed. Concrete masonry units come in a variety of shapes, sizes, and finishes but are commonly found in 4-inch (102-mm) or 8-inch (203-mm) increments; 8 inches (203 mm) high by 8 inches (203 mm) wide by 16 inches (406 mm) long for example. This characteristic makes it relatively easy to compute dimensions when using masonry. Masonry can be used to form footings and foundations for crawl spaces as well as basements. CMUs are used in various areas throughout the United States in residential construction.

In addition to the basic requirements for footings and foundations, the IRC also provides criteria to deal with unusual soil conditions and types. As we have seen previously, the IRC allows for a wide variety of soil types that can be encountered in different regions of the country. Presumptive load bearing values for different types of soil conditions are provided for in Table R401.4.1, as shown in Figure 6-3. Where unusual soil conditions occur, the IRC authorizes the code official to require a soil study and tests to determine the specific characteristics present in the soil. Where soil types are particularly ill suited to support structures, she may require an engineered design to be provided that takes into account the specific conditions in the soil.

Finally, Chapter 4 provides criteria for the proper treatment of **under-floor areas** (crawl spaces) in dwellings. Here, criteria are found to address the proper ventilation of under-floor areas including provisions for both vented and unvented crawl space conditions. These IRC provisions do not apply to under-floor spaces that are classified as basements. The required ventilation area is established at the ratio of 1 square foot (0.093 m²) of ventilation for every 150 square feet (13.9 m²) of under-floor space, and one vent is required within 3 feet (914 mm) of every corner of the under-floor area.

Minimum access requirements are provided for under-floor areas as well. Here, the code identifies a minimum opening size in the floor or adjacent side wall to allow access for repairs and maintenance (see Figure 6-4). The IRC requires under-floor areas to be free of vegetation and construction debris prior to occupancy. The finished grade under the floor area may be left at the same level as the bottom of the footings unless groundwater is present or where surface water does not drain from the site. In these cases, the grade height under the floor must match the grade outside the under-floor area or an approved drainage system must be provided.

Figure 6-3

Table R401.4.1 Presumptive load-bearing values of foundation materials.

TABLE R401.4.1
PRESUMPTIVE LOAD-BEARING VALUES OF
FOUNDATION MATERIALS[a]

CLASS OF MATERIAL	LOAD-BEARING PRESSURE (pounds per square foot)
Crystalline bedrock	12,000
Sedimentary and foliated rock	4,000
Sandy gravel and/or gravel (GW and GP)	3,000
Sand, silty sand, clayey sand, silty gravel and clayey gravel (SW, SP, SM, SC, GM and GC)	2,000
Clay, sandy clay, silty clay, clayey silt, silt and sandy silt (CL, ML, MH and CH)	1,500[b]

For SI: 1 pound per square foot = 0.0479 kPa.

a. When soil tests are required by Section R401.4, the allowable bearing capacities of the soil shall be part of the recommendations.

b. Where the building official determines that in-place soils with an allowable bearing capacity of less than 1,500 psf are likely to be present at the site, the allowable bearing capacity shall be determined by a soils investigation.

Figure 6-4

A crawl space access hole located under a stairway, minimum required size 18 inches by 24 inches (457 mm by 610 mm).

COURTESY DELMAR/CENGAGE LEARNING

GREEN OPTIONS

At first glance, it would appear that there are few green options available when dealing with foundations. After all, there seem to be relatively few choices to be made here: wood, concrete, or masonry. Upon closer examination, however, there are a number of green options that affect foundations directly, as well as some green decisions that affect the site.

Another factor affecting green decisions has to do with which particular aspects of green construction and sustainable development are most important to the individual making the decisions. In the case of footings and foundations, the decisions one makes or the course of action one takes is largely dependent on one's opinion of what makes something green in the first place.

For example, one who feels that the construction site and surrounding properties should be protected and developed with minimal impact might take a different path than someone who feels green means to use recycled materials where possible. It is, as they say, a matter of perspective. The following sections will explore green options that exist related to foundations, taking into account some of these considerations.

SITE ISSUES

Because foundation construction occurs at the interface between the construction and the earth, it stands to reason that the preparations necessary to place a foundation (excavation, grading, erosion control) have the potential to impact the site in a profound way. For example, surface runoff into streams and wetlands can cause pollution and **turbidity** (cloudiness or murkiness due to sediments suspended in water), which are often cause for concern. Grading spoils can be disposed of improperly as well, all of which can adversely affect the environment if not handled responsibly.

Although the IRC does not contain code language that directly regulates these topics, many jurisdictions adopt local ordinances that do regulate them, so it is important to understand how they impact the preparatory work and placement of a foundation on a dwelling site. It stands to reason that there is homework to be done here. Some jurisdictions adopt Appendix J, Grading from the International Building Code, and use it as the basis for the regulation of excavations, fills, grading, and erosion control. Some jurisdictions may also require a permit to be issued prior to performing any earthwork. The responsible handling of each of these areas is decidedly green and has the potential to impact the project in a positive way. Let's explore each topic more fully.

Excavation and Grading

It is often necessary to change the surface grade when preparing a site to receive a foundation. In some cases, for example, **excavation** (digging into the earth) is necessary to prepare the site for a basement. In other cases, the general terrain of the site must be modified to better accommodate the size and shape of the dwelling or to create a more usable site, which is referred to as **grading**. At the very least and regardless of which must be done to prepare the site, it is necessary to dig through the organic material into what is referred to as "native," undisturbed soil suitable for supporting a foundation as required by IRC Section R403.1.

In any case, some movement of the earth is almost always required to prepare the site; occasionally, it is sometimes necessary to move very large quantities of soil. This earthwork can impact the environment in many ways. Where large excavations occur, the quantities of earth removed can be quite large. In fact, they can be so large that they cannot conveniently be disposed of on the actual construction site, making it necessary to transport all or a portion of the soil to another location. When this occurs, the site to which the earth is transported can be adversely affected if the spoils from the excavation are dumped irresponsibly. Ideally, the excess earth will be placed in a location where it is needed for some other purpose or placed in a manner that does not adversely affect the target site, such as if it were dumped into a protected wetland.

In contrast, it is sometimes necessary to import soil or **granular fill** (crushed rock, gravel, and similar) to level a site or to otherwise prepare it for construction. Sometimes the fill is necessary to make the site more usable, say to create flat area for a yard, lawn, or patio and in some cases it is even necessary to use fill to support the dwelling. Where fill is imported to a site, it is necessary to place it in a responsible way, so as not to adversely affect the site or neighboring properties.

TAKE NOTE!

Where it is necessary to use fill to support a dwelling (structural fill), Section R403.1 of the IRC requires it to be engineered. This means that the material used must be evaluated by a qualified engineer or a person with credentials acceptable to the building official and deemed suitable for use as a fill material. The placement of the fill must be carefully monitored and documented by the design professional as well. It is typically placed in lifts 8 inches (203 mm) to 12 inches (305 mm) thick and rolled or compacted to ensure that it will not settle excessively. The jurisdiction will almost certainly require a fill certification report to be submitted, on which the engineer documents that the fill is suitable to support the dwelling.

Why It's Green

Utilizing the natural terrain of a site to the maximum extent possible is green because it minimizes the amount of excavation and grading necessary to develop the site. This, in turn, can reduce impact to natural drainage ways and areas that are environmentally sensitive such as wetlands, streams, and fish corridors. It also reduces the amount of soil that must be relocated or removed from the site, which is costly and consumes large amounts of energy. Preserving tree canopy to the maximum extent feasible is also green because trees provide shade for dwellings which can lower energy costs during summer and winter.

If the fills are placed in such a way that they modify a natural drainage way and divert water to a neighbor's property or into an environmentally sensitive area, the results can be devastating and sometimes irreversible. Consult with the local regulatory agencies and professionals such as geotechnical engineers to ensure that fills are placed in a safe, legal, and responsible manner.

Intent

The intent of these provisions, where adopted, is to ensure that earthwork such as excavation and grading is conducted safely and in a way that minimizes impacts to adjacent properties. These regulations further intend that site conditions to remain after the earthwork is completed are stable and will provide adequate drainage from the property.

Green Decisions and Limitations

What can I do to minimize the impact that the site preparation for my green dwelling will have on the environment?

Are there laws or local ordinances that impact this decision?

Options

Where excavation, grading, or fill are required to prepare a building site for construction, consider the following green options.

Utilize the natural terrain: One of the most effective ways to minimize the impact that earthwork has on the environment is to utilize the natural terrain to the maximum extent possible. If, for example, a portion of a sloping site would accommodate a **daylight basement** (a basement with at least one wall fully above grade) rather than a traditional basement, it might take considerably less excavation to accomplish the daylight basement (see Figure 6-5). Where a site has natural topographic features such as a natural ridge or level area, it makes sense that these areas might be more easily adapted for the dwelling site than, say, another area on the site that would take a large amount of fill to make it suitable for construction.

In other cases, the natural topography of the site might make an earth-sheltered dwelling a viable option for all or part of the structure. In any case, it is not bad advice to utilize the natural strengths and features of the site to the fullest advantage.

Erosion control: Some areas, like in the northwest region of the United States, actually require erosion control measures to be utilized during the site preparation and construction process to minimize the impact to streams and fish habitat. During construction, silts and sediments can be washed into waterways, adversely affecting them. Erosion control measures such as silt fencing and the use

Figure 6-5

Sunlight streaming in through a daylight basement wall. Where practical, daylight basements can help utilize sloping sites to the fullest advantage.

COURTESY DELMAR/CENGAGE LEARNING

Figure 6-6

A silt fence installed along the edge of a property for erosion control.

COURTESY DELMAR/CENGAGE LEARNING

of straw to cover the ground during construction can greatly reduce the amount of sediment that washes into adjacent waterways (see Figure 6-6).

Even where erosion control measures are not mandated by a local authority, the possibility of using them should be considered where construction occurs adjacent to streams, creeks, rivers, and wetlands. Effective erosion control is often quickly and easily accomplished and can have a profoundly positive effect on the environment.

Section R401.3 Drainage

Surface drainage is regulated by the provisions of this IRC code section. The charging statement requires all surface drainage from a construction site to be diverted to a storm sewer or other approved point of collection. Generally, the IRC requires a dwelling site to be graded such that the final grade slopes away from the dwelling rather than toward it. This is to ensure that water is safely channeled away from the building, which helps to prevent flooded crawl spaces and basements.

The IRC requires the final grade to fall a minimum of 6 inches (152 mm) within the first 10 feet (3,048 mm) of the structure. Where this requirement cannot be met, say, due to the particular topographic conditions on the site or where the dwelling is less than 10 feet (3,048 mm) from a property line, the IRC requires the grade to slope away from the building at a minimum slope of 5%, which is 0.5 inch (13 mm) of fall for every 12 inches (305 mm) of horizontal distance. The code also requires that swales or drains be used where necessary to ensure drainage away from the structure.

Why It's Green

Where there are **impervious surfaces** (which do not absorb water like concrete and asphalt) within 10 feet (3,048 mm) of the structure, these surfaces must slope away at a 2% grade, which is 0.125 inch (3 mm) of fall for every 12 inches (305 mm) of horizontal distance.

Intent

The intent of this IRC section is to ensure that surface water is effectively channeled away from the dwelling. The code further intends that surface water be disposed of in an approved manner. The IRC also intends that where there are impervious surfaces adjacent to a dwelling, they also slope away from the structure to further facilitate drainage.

Green Decisions and Limitations

Shall I store some or all of the surface drainage onsite?
Is pervious concrete an option instead of traditional concrete?
How will local rules and ordinances impact this decision?

Options

When selecting a method of surface drainage to satisfy the requirements of IRC Section R401.3, consider the following green alternatives.

Onsite storage: Where local rules and site conditions allow, it may be possible to store all or some of the surface drainage on the dwelling site. Incorporating water features such as ponds and natural waterscapes into the design of the site to collect surface drainage can reduce demands placed on local infrastructure. Use of native plants and aquatic vegetation can help to filter water of contaminants before it makes its way back into waterways or ground water. Some cities, such as Portland, Oregon, offer incentives to encourage its citizens to disconnect their rain drains from the municipal system. Check with local authorities to ensure that this practice is allowed. Also, onsite storage of surface drainage must be carefully designed to ensure that the storage area is large enough to accommodate the flows and also so that overflows are handled in a code-compliant way. Again, check with the local jurisdiction to ensure that onsite storage of surface drainage is done legally.

Pervious concrete: Pervious concrete, as the name implies, is a special type of concrete that allows water to soak through rather than to be shed off of the surface as is the case with the traditional material. Thus, it might be used for paved areas such as driveways, sidewalks, and in other nonstructural applications. Pervious concrete has the advantage of reducing surface runoff which, as we have already discussed, has many environmental benefits. Although Section R105.2 of the IRC lists sidewalks and driveways as exempt, many jurisdictions regulate public sidewalks and other concrete work in the public right-of-way under different regulations. In these cases, pervious concrete may not be allowed and the reader is advised to check with the local jurisdiction for specific requirements and prohibitions.

Section R403.1.7 Footings On or Adjacent to Slopes

The IRC requires footings used to support structures to be located a safe distance from slopes that are steeper than 1 unit vertical to 3 units horizontal (1V : 3H). This is true for dwellings constructed at both the top and the bottom of the slope, although the method by which the safe distance is computed differs depending on conditions. These clearances are established to ensure that the placement of a dwelling near a slope does not have an adverse effect on the slopes in an effort to keep them stable. Slopes that are flatter than 1V : 3H are relatively stable and are less affected by construction than their steeper counterparts (see Figure 6-7).

Foundations located at the top of slopes require special consideration. Previously it was mentioned that all forces within a structure are ultimately transmitted to the foundation. Once the forces are transmitted to the footings, they are then transferred into the earth. Without getting too technical, load transfer from the footings occurs at roughly a 45 degree angle from the edge of the footing into the soil. Where footings are located too closely to the top of a slope, they can potentially cause portions of the slope to "shear" off as a result, causing a serious and potentially unstable situation. Safe distances from the top of a slope are a function of the total height of the slope, again as seen in Figure 6-7.

Foundations are also required to be a minimum distance from the bottom or the "toe" of a slope. These minimum distances are established to prevent the excavations and site preparation work from impacting the toe of the slope, again causing an unstable situation that could potentially destabilize the hillside causing a landslide. This minimum distance is also established to allow space from the front of the structure to the slope should the slope move on its own. The minimum distance from the toe of the slope is also a function of height, as seen in Figure 6-7. Alternate setbacks to slopes are allowed when approved by the building official, provided it can be demonstrated through an engineering study that the slopes will remain stable once the work is performed. Such studies are often referred to as geotechnical engineering reports.

Figure 6-7

Figure R403.1.7.1
Foundation clearance
from slopes.

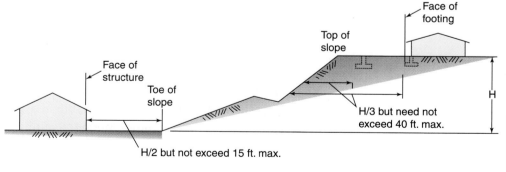

For SI: 1 foot = 304.8 mm

COPYRIGHT 2009 IRC®

Why It's Green

Locating a dwelling a safe distance from the head or toe of a slope is a sustainable practice because it ensures that the site will remain stable during and after construction. Regardless of how "green" a dwelling is constructed, it is potentially lost if built on an unstable site.

Intent

The intent of this IRC section is to establish safe clearances for foundations from slopes to avoid the potential for landslide. The IRC further intends that foundations be located a safe distance from slopes to avoid the risk of damage to the structure should a slide occur on its own.

Green Decisions and Limitations

Where shall I locate my green dwelling relative to the slope on my building site?

How does the IRC impact this decision?

Reduced Clearances Option

In a perfect world, it would always be possible to locate a dwelling with the clearances outlined in Section R403.1.7. However, in reality, these clearances are not always attainable. In these cases, the building official may grant alternate clearances from slopes as permitted under the provisions of Section R403.1.7.4, provided an investigation can be conducted that will satisfactorily demonstrate that the site is stable despite the reduced clearances. The cost of such a geotechnical investigation is justified where such unusual site conditions exist.

Concrete Footings, Foundations, and Retaining Walls

The IRC provides material standards and prescriptive code requirements for concrete and masonry footings and foundations. The minimum compressive strength for concrete is established in Table R402.2, which varies based on the particular application in which the concrete is used and the specific weathering potential present for the region (see Figure 6-8). Once again, this table is a great example of the IRC's flexibility, as it covers virtually all conditions under which concrete is used.

The American Concrete Institute (ACI) Standard ACI 318, Building Code Requirements for Structural Concrete, is referenced throughout the text. Materials used in the preparation and testing of concrete must comply with Chapter 3 of ACI 318. Additionally, ACI 332, Requirements of Residential Concrete Construction, is also referenced, as are various ASTM standards. Footing widths are defined in Table R403.1 and are primarily a function of the soil load bearing values derived from Table R401.4.1 (see Figure 6-3) and the number of stories to be supported. It stands to reason that the greater the number of stories supported, the wider the footings need to be to safely support the weight of the structure (see Figure 6-9).

Requirements for foundation walls are also established in Chapter 4. Foundation walls vary in width and height depending upon the application in which they are needed. For example, a standard foundation wall 2 feet (610 mm) in height must support less soil than a foundation wall that forms a full height basement. As the height of the earth fill increases, so must either the thickness of the wall or the quantity of reinforcing steel increase to support the increased load. Table R404.1.1(2), in Figure 6-10, shows the various reinforcing steel requirements for an 8-inch

Figure 6-8

Table R402.2 Minimum specified compressive strength of concrete.

TABLE R402.2
MINIMUM SPECIFIED COMPRESSIVE STRENGTH OF CONCRETE

TYPE OR LOCATION OF CONCRETE CONSTRUCTION	MINIMUM SPECIFIED COMPRESSIVE STRENGTH[a] (f'_c)		
	Weathering Potential[b]		
	Negligible	Moderate	Severe
Basement walls, foundations and other concrete not exposed to the weather	2,500	2,500	2,500[c]
Basement slabs and interior slabs on grade, except garage floor slabs	2,500	2,500	2,500[c]
Basement walls, foundation walls, exterior walls and other vertical concrete work exposed to the weather	2,500	3,000[d]	3,000[d]
Porches, carport slabs and steps exposed to the weather, and garage floor slabs	2,500	3,000[d, e, f]	3,500[d, e, f]

For SI: 1 pound per square inch = 6.895 kPa.

a. Strength at 28 days psi.

b. See Table R301.2(1) for weathering potential.

c. Concrete in these locations that may be subject to freezing and thawing during construction shall be air-entrained concrete in accordance with Footnote d.

d. Concrete shall be air-entrained. Total air content (percent by volume of concrete) shall be not less than 5 percent or more than 7 percent.

e. See Section R402.2 for maximum cementitious materials content.

f. For garage floors with a steel troweled finish, reduction of the total air content (percent by volume of concrete) to not less than 3 percent is permitted if the specified compressive strength of the concrete is increased to not less than 4,000 psi.

Figure 6-9

Table R403.1 Minimum width of concrete, precast, or masonry footings.

TABLE R403.1
MINIMUM WIDTH OF CONCRETE,
PRECAST OR MASONRY FOOTINGS
(inches)[a]

	LOAD-BEARING VALUE OF SOIL (psf)			
	1,500	2,000	3,000	≥ 4,000
Conventional light-frame construction				
1-story	12	12	12	12
2-story	15	12	12	12
3-story	23	17	12	12
4-inch brick veneer over light frame or 8-inch hollow concrete masonry				
1-story	12	12	12	12
2-story	21	16	12	12
3-story	32	24	16	12
8-inch solid or fully grouted masonry				
1-story	16	12	12	12
2-story	29	21	14	12
3-story	42	32	21	16

For SI: 1 inch = 25.4 mm, 1 pound per square foot = 0.0479 kPa.

a. Where minimum footing width is 12 inches, use of a single wythe of solid or fully grouted 12-inch nominal concrete masonry units is permitted.

(203-mm) thick masonry wall. Note how the size and spacing of the rebar changes as a function of the height of the wall.

Intent

The intent of these IRC sections is to ensure that appropriate steps be taken in the preparation of the site. These provisions also ensure that a foundation will be safely located in relationship to hazards and that it will drain surface water in an approved manner. The code further establishes minimum standards for concrete and masonry used to form footings and foundation walls. Additionally, the IRC intends that footings and foundation walls be constructed in accordance with the prescriptive width, thickness, height, and reinforcing steel requirements prescribed in the code to ensure that structures built on top of them are safely supported or that they are engineered for given loading conditions.

Figure 6-10

Table R404.1.1(2) 8-inch masonry foundation walls with reinforcing where $d > 5$ inches.

TABLE R404.1.1(2)
8-INCH MASONRY FOUNDATION WALLS WITH REINFORCING
WHERE d > 5 INCHES[a, c]

WALL HEIGHT	HEIGHT OF UNBALANCED BACKFILL[e]	MINIMUM VERTICAL REINFORCEMENT AND SPACING (INCHES)[b, c]		
		Soil classes and lateral soil load[d] (psf per foot below grade)		
		GW, GP, SW and SP soils 30	GM, GC, SM, SM-SC and ML soils 45	SC, ML-CL and inorganic CL soils 60
6 feet 8 inches	4 feet (or less)	#4 at 48	#4 at 48	#4 at 48
	5 feet	#4 at 48	#4 at 48	#4 at 48
	6 feet 8 inches	#4 at 48	#5 at 48	#6 at 48
7 feet 4 inches	4 feet (or less)	#4 at 48	#4 at 48	#4 at 48
	5 feet	#4 at 48	#4 at 48	#4 at 48
	6 feet	#4 at 48	#5 at 48	#5 at 48
	7 feet 4 inches	#5 at 48	#6 at 48	#6 at 40
8 feet	4 feet (or less)	#4 at 48	#4 at 48	#4 at 48
	5 feet	#4 at 48	#4 at 48	#4 at 48
	6 feet	#4 at 48	#5 at 48	#5 at 48
	7 feet	#5 at 48	#6 at 48	#6 at 40
	8 feet	#5 at 48	#6 at 48	#6 at 32
8 feet 8 inches	4 feet (or less)	#4 at 48	#4 at 48	#4 at 48
	5 feet	#4 at 48	#4 at 48	#5 at 48
	6 feet	#4 at 48	#5 at 48	#6 at 48
	7 feet	#5 at 48	#6 at 48	#6 at 40
	8 feet 8 inches	#6 at 48	#6 at 32	#6 at 24
9 feet 4 inches	4 feet (or less)	#4 at 48	#4 at 48	#4 at 48
	5 feet	#4 at 48	#4 at 48	#5 at 48
	6 feet	#4 at 48	#5 at 48	#6 at 48
	7 feet	#5 at 48	#6 at 48	#6 at 40
	8 feet	#6 at 48	#6 at 40	#6 at 24
	9 feet 4 inches	#6 at 40	#6 at 24	#6 at 16
10 feet	4 feet (or less)	#4 at 48	#4 at 48	#4 at 48
	5 feet	#4 at 48	#4 at 48	#5 at 48
	6 feet	#4 at 48	#5 at 48	#6 at 48
	7 feet	#5 at 48	#6 at 48	#6 at 32
	8 feet	#6 at 48	#6 at 32	#6 at 24
	9 feet	#6 at 40	#6 at 24	#6 at 16
	10 feet	#6 at 32	#6 at 16	#6 at 16

For SI: 1 inch = 25.4 mm, 1 foot = 304.8 mm, 1 pound per square foot per foot = 0.157 kPa/mm.

a. Mortar shall be Type M or S and masonry shall be laid in running bond.

b. Alternative reinforcing bar sizes and spacings having an equivalent cross-sectional area of reinforcement per lineal foot of wall shall be permitted provided the spacing of the reinforcement does not exceed 72 inches.

c. Vertical reinforcement shall be Grade 60 minimum. The distance, d, from the face of the soil side of the wall to the center of vertical reinforcement shall be at least 5 inches.

d. Soil classes are in accordance with the Unified Soil Classification System and design lateral soil loads are for moist conditions without hydrostatic pressure. Refer to Table R405.1.

e. Unbalanced backfill height is the difference in height between the exterior finish ground level and the lower of the top of the concrete footing that supports the foundation wall or the interior finish ground level. Where an interior concrete slab-on-grade is provided and is in contact with the interior surface of the foundation wall, measurement of the unbalanced backfill height from the exterior finish ground level to the top of the interior concrete slab is permitted.

Green Decisions and Limitations

Can I use recycled concrete in the construction of my green dwelling?

What options exist for insulating my foundation, basement, and crawl space?

Options

When making decisions regarding materials and construction methods for concrete footings, foundations and retaining walls, consider the following green options.

Recycled concrete used as aggregate: Concrete is one of the most recycled materials around. According to the Portland Cement Association (PCA), 38 states use recycled concrete as an aggregate base course (under roads and driveways) and 11 states recycle concrete into new concrete. It is reported that concrete manufactured with recycled materials performs the same as that made with standard aggregates (see Figure 6-11). Recycled concrete can readily be incorporated into new concrete or obtained for use as a base course in many areas of the United States. Check with a local concrete supplier for availability. ACI 555 provides criteria for processing recycled concrete into aggregate. Be sure that the recycled aggregates, where available, comply with this standard.

Recycled concrete retaining wall blocks: Recycled concrete is used in the manufacture of a number of concrete-based products. Some concrete is simply broken into large pieces and made into retaining walls or utilized as steps and patio pavers such as those seen in Figure 6-12. Recycled concrete is also used in the manufacture of patio paving blocks, fountains, and other landscape products. Many home and garden centers now carry concrete-based building materials and landscaping products made from recycled concrete. Recall from Chapter 5 that deconstruction is green because it allows for maximum amounts of construction waste materials to be recycled. A local concrete recycling center is a great way to dispose of concrete waste from a green dwelling project.

Why It's Green

Recycled concrete is green because it utilizes waste materials by crushing it and using it again as aggregate in new concrete. Recycled concrete is also used to form modular retaining walls, patio pavers, and other landscape materials. Insulating concrete forms (ICF) are green because they provide not only the basic form work for the placement of concrete, but also extra insulation from the solid foam plastic insulation that remains in place after the concrete is poured.

Figure 6-11

Recycled concrete crushed and made into aggregate.

COURTESY OF PORTLAND CEMENT ASSOCIATION

Figure 6-12
A beautiful patio, retaining wall, and steps made of recycled concrete salvaged from a demolished basketball court. This creative use of scrap material transforms an otherwise unusable sloping portion of the yard into a lovely seating area.

COURTESY OF J BREW

TAKE NOTE!

According to IRC Section R105.2, permits are required for all retaining walls taller than 4 feet (1,219 mm) in height measured from the bottom of the footing to the top of the wall. Permits are also required where the retaining wall supports a **surcharge** (additional loads other than level soil adjacent to the wall, such as hillside slopes) regardless of height. Be sure to check with the local jurisdiction for permit requirements in your area.

Insulating concrete forms: These are increasingly popular for use in foundations, particularly where they form basement walls and heated crawl space areas. ICF offer higher insulating values than plain concrete walls and offer a modular concrete form all in one convenient package. ICF products are available in a wide variety of sizes and types. Most of the ICF products available today have ICC Evaluation Reports or Legacy Reports available which govern how the products are used and their limitations. Additionally, the IRC provides prescriptive code criteria for these products, making them easy to incorporate into the green dwelling. More on ICF in Chapter 8.

Post and beam foundations: Where environmental or other concerns arise around the use of concrete or where the use of conventional footings and foundations are impractical, post and beam foundations provide an effective solution.

Post and beam foundations can use considerably less concrete than conventional footings and foundations because, in some cases, the only concrete used is in the intermittent footing locations to which the supporting posts are attached. Post and beam foundations may consist of wood posts and beams combined with concrete piers, concrete or steel piles combined with concrete grade beams, or other configurations. The types of foundations are effective for steeply sloped lots or in cases where the construction of conventional spread-type footings is otherwise not desired. Remember, post and beam type foundations are not prescriptive so there are no provisions contained in the IRC for their use. Thus, engineering will be required.

CHAPTER 6 APPLIED TO THE SAMPLE PROJECT

Chapter 6 addressed matters related to the requirements for foundations. Additionally, Chapter 6 addressed code requirements for under-floor areas, excavation and grading, construction adjacent to slopes, and more. The following analysis will demonstrate the application of some of the IRC provisions for foundations with consideration given for the proposed scope of work and our design philosophy.

Site Considerations

In this example, we are told that the site is generally flat, so it is a safe assumption that minimal grading will be involved. The IRC requires that footings be founded on native, undisturbed soil, so we know that at the very least we will have to excavate to the required frost depth as established by the jurisdiction. Local regulations may require erosion control measures, so we will check with the local jurisdiction for requirements. Given the flat site conditions, it will be relatively easy to meet the grading requirements of IRC Section R401.3. Additionally, we will select pervious paving for those nonstructural applications such as sidewalks and driveways, assuming these materials are acceptable to the jurisdiction because this is consistent with our desire to minimize disruption to the site.

Building Considerations

The project scope calls for a slab foundation which will require engineering due to the use of rammed-earth walls. Remember that the prescriptive footing and foundation criterion found in the IRC applies only to conventional construction. Given that the slab and footings must support the dead loads from the massive rammed-earth walls, plus the weights of the roof and other building systems in addition to all wind and seismic loads as the case may be, the design professional will have to carefully consider all these factors in her design. The foundation design must comply with American Concrete Institute Standard ACI 318. The 3-foot (914-mm) proposed clearance from the property line requires no special consideration for the foundation.

CHAPTER SUMMARY

Chapter 4 of the IRC covers foundations for residential construction. In addition to specific requirements for footings and foundations, Chapter 4 also covers requirements related to the design of the site. Following are highlights of the chapter.

- Chapter 4 provides specific requirements for footings and foundations through the use of text, tables, and figures.
- Minimum footing thicknesses and widths are prescribed within the provisions of the chapter.
- Minimum foundation wall thicknesses and reinforcing steel requirements are provided for concrete and masonry walls.
- Assumed soil load bearing values are provided for all major soil types.
- Site drainage requirements are established.
- Slope clearance requirements are provided for foundations located at both the top and bottom of hillsides.
- Several green options exist related to the preparation of the site (such as with grading and excavation) but are not specifically identified as such in the IRC.
- Other green options exist for the use of recycled concrete, recycled concrete aggregate, and post and beam type foundations.

FLOOR FRAMING

This chapter corresponds to IRC Chapter 5, Floors

learning objective

To know and understand how floors and floor framing systems are regulated under the provisions of the IRC and to explore options for green floor construction.

IRC CHAPTER 5—OVERVIEW

Chapter 5 of the International Residential Code regulates the construction of floors for dwellings. Floors provide the primary living and walking surfaces within a dwelling on which we sleep, eat, and play. More than that, floors form an important part of the structural system that supports a dwelling, distributing forces to the supporting walls which, in turn, distribute those forces into another floor or to the foundation. The IRC establishes code criteria for floors to ensure that loads are safely and effectively transferred to the resisting elements. The IRC also establishes criteria for nonstructural floors (i.e., slab on grade) and spaces under floors, such as crawl spaces.

Floors are made up of horizontal structural members called joists (see Figure 7-1). The ends of joists are typically supported by walls but can also be supported by beams, ledgers, and other structural elements within the building. Floor joists support two primary types of load: gravity loads and lateral forces. Gravity loads act on floors at all times and cause these horizontal framing members to bend. As the joists bend in response to the weight of people, appliances, furnishings, and the like, their strengths (i.e., their structural properties) come into play and determine, in part, how far the joist can span and how much weight it can support under given loading conditions.

How far a joist can span is a function of several important factors. The IRC establishes the specific design criteria for different room types in Table R301.5 of the code (see Chapter 5 of this text for a review). In bedrooms, for example, the IRC requires a basic floor live loading requirement of 30 pounds per square foot (1.437 kPa) of floor area as the basic design criterion. In other living areas, the basic load requirement is 40 pounds per square foot (1.92 kPa) (see Figure 7-2).

Figure 7-1

Conventional wood floor framing installed in a dwelling. In this case, the ends of the joists are supported by a glued laminated beam.

COURTESY DELMAR/CENGAGE LEARNING

remember

Recall from Chapter 1 of this text that gravity loads are those ever-present forces generated by the weights of the construction materials themselves (dead load) plus the variable weights of the contents of the dwelling (live load). Lateral forces are those intermittent horizontal and sometimes vertical forces generated by the wind and seismic loads.

To put this into perspective, picture the following: Within a dwelling is a bedroom with the dimensions of 10 feet (3,048 mm) by 10 feet (3,048 mm) or 100 square feet (9.29 m²) in area. Under the criteria established in the code, for every square foot of floor area in the bedroom, the floor joists are required to support 30 pounds (1.437 kPa) of live load or a total weight of 3,000 pounds (1,361 kg), in addition to the weight of the floor framing itself! For simplicity's sake, the IRC assumes the required 3,000 pound (1,361 kg) total load is uniformly distributed.

Joist spans are also a function of the particular species and grade of the lumber material. Some species of lumber have structural properties that are better suited than others and are thus better suited for structural applications. In the west, Douglas fir is commonly used in construction. In other areas of the United States, southern yellow pine is used. Largely, the type of lumber used in construction is a matter of what is available locally.

Figure 7-2

Table R301.5 Minimum uniformly distributed live loads.

TABLE R301.5
MINIMUM UNIFORMLY DISTRIBUTED LIVE LOADS
(in pounds per square foot)

USE	LIVE LOAD
Attics without storage[b]	10
Attics with limited storage[b, g]	20
Habitable attics and attics served with fixed stairs	30
Balconies (exterior) and decks[e]	40
Fire escapes	40
Guardrails and handrails[d]	200[h]
Guardrail in-fill components[f]	50[h]
Passenger vehicle garages[a]	50[a]
Rooms other than sleeping room	40
Sleeping rooms	30
Stairs	40[c]

For SI: 1 pound per square foot = 0.0479 kPa, 1 square inch = 645 mm², 1 pound = 4.45 N.

a. Elevated garage floors shall be capable of supporting a 2,000-pound load applied over a 20-square-inch area.

b. Attics without storage are those where the maximum clear height between joist and rafter is less than 42 inches, or where there are not two or more adjacent trusses with the same web configuration capable of containing a rectangle 42 inches high by 2 feet wide, or greater, located within the plane of the truss. For attics without storage, this live load need not be assumed to act concurrently with any other live load requirements.

c. Individual stair treads shall be designed for the uniformly distributed live load or a 300-pound concentrated load acting over an area of 4 square inches, whichever produces the greater stresses.

d. A single concentrated load applied in any direction at any point along the top.

e. See Section R502.2.2 for decks attached to exterior walls.

f. Guard in-fill components (all those except the handrail), balusters and panel fillers shall be designed to withstand a horizontally applied normal load of 50 pounds on an area equal to 1 square foot. This load need not be assumed to act concurrently with any other live load requirement.

g. For attics with limited storage and constructed with trusses, this live load need be applied only to those portions of the bottom chord where there are two or more adjacent trusses with the same web configuration capable of containing a rectangle 42 inches high or greater by 2 feet wide or greater, located within the plane of the truss. The rectangle shall fit between the top of the bottom chord and the bottom of any other truss member, provided that each of the following criteria is met.
  1. The attic area is accessible by a pull-down stairway or framed in accordance with Section R807.1.
  2. The truss has a bottom chord pitch less than 2:12.
  3. Required insulation depth is less than the bottom chord member depth.
  The bottom chords of trusses meeting the above criteria for limited storage shall be designed for the greater of the actual imposed dead load or 10 psf, uniformly distributed over the entire span.

h. Glazing used in handrail assemblies and guards shall be designed with a safety factor of 4. The safety factor shall be applied to each of the concentrated loads applied to the top of the rail, and to the load on the in-fill components. These loads shall be determined independent of one another, and loads are assumed not to occur with any other live load.

COPYRIGHT 2009 IRC®

The final consideration in determining the span of floor joists is the spacing of the framing material. It stands to reason that the more closely spaced the floor joists are to one another, the less load each individual joist must carry. Floor joists are commonly spaced on 16-inch (406-mm) centers, but 12-inch (305-mm) centers are also common. To a lesser degree, a spacing of 24-inch (610-mm) centers is used. So, the loading criteria, species and grade, and spacing of the joists must all be taken into account to determine how far the joist can safely span.

Fortunately, the IRC makes this easy through the use of tables, as shown in Figure 7-3. Table R502.3.1(1) shows safe joist spans for a variety of species and grades of wood joists under various spacing conditions. Note that the heading of this table identifies that it is to be used for sleeping room loading conditions. The use of this table would not be allowed for loading conditions greater than 30 pounds per square foot (1.437 kPa), such as in a dining room.

There is another important criterion to be considered when discussing floor joists: deflection. The amount of **deflection** (actual amount of movement of the joist under given load) is also regulated by the IRC. The amount of deflection is especially important for floors because the finish materials attached to them (e.g., ceramic floor tile on top of the floor or gypsum board applied to ceilings below) can only bend so far without cracking. The deflection limitations help these materials to perform as intended.

Human comfort is another important factor considered in the deflection criteria. Floors that deflect excessively can be bouncy and feel less "sound," which can be unnerving for some, as humans can be highly sensitive to such movement. The table in Figure 7-3 takes deflection into account.

Figure 7-3 Table R502.3.1(1) Floor joist spans for common lumber species (residential sleeping areas, live load = 30 psf).

TABLE R502.3.1(1)
FLOOR JOIST SPANS FOR COMMON LUMBER SPECIES
(Residential sleeping areas, live load = 30 psf, L/Δ = 360)[a]

JOIST SPACING (inches)	SPECIES AND GRADE		DEAD LOAD = 10 psf				DEAD LOAD = 20 psf			
			2 × 6	2 × 8	2 × 10	2 × 12	2 × 6	2 × 8	2 × 10	2 × 12
			Maximum floor joist spans							
			(ft - in.)	(ft - in.)	(ft - in.)	(ft - in.)	(ft - in.)	(ft - in.)	(ft - in.)	(ft - in.)
12	Douglas fir-larch	SS	12-6	16-6	21-0	25-7	12-6	16-6	21-0	25-7
	Douglas fir-larch	#1	12-0	15-10	20-3	24-8	12-0	15-7	19-0	22-0
	Douglas fir-larch	#2	11-10	15-7	19-10	23-0	11-6	14-7	17-9	20-7
	Douglas fir-larch	#3	9-8	12-4	15-0	17-5	8-8	11-0	13-5	15-7
	Hem-fir	SS	11-10	15-7	19-10	24-2	11-10	15-7	19-10	24-2
	Hem-fir	#1	11-7	15-3	19-5	23-7	11-7	15-2	18-6	21-6
	Hem-fir	#2	11-0	14-6	18-6	22-6	11-0	14-4	17-6	20-4
	Hem-fir	#3	9-8	12-4	15-0	17-5	8-8	11-0	13-5	15-7
	Southern pine	SS	12-3	16-2	20-8	25-1	12-3	16-2	20-8	25-1
	Southern pine	#1	12-0	15-10	20-3	24-8	12-0	15-10	20-3	24-8
	Southern pine	#2	11-10	15-7	19-10	24-2	11-10	15-7	18-7	21-9
	Southern pine	#3	10-5	13-3	15-8	18-8	9-4	11-11	14-0	16-8
	Spruce-pine-fir	SS	11-7	15-3	19-5	23-7	11-7	15-3	19-5	23-7
	Spruce-pine-fir	#1	11-3	14-11	19-0	23-0	11-3	14-7	17-9	20-7
	Spruce-pine-fir	#2	11-3	14-11	19-0	23-0	11-3	14-7	17-9	20-7
	Spruce-pine-fir	#3	9-8	12-4	15-0	17-5	8-8	11-0	13-5	15-7
16	Douglas fir-larch	SS	11-4	15-0	19-1	23-3	11-4	15-0	19-1	23-0
	Douglas fir-larch	#1	10-11	14-5	18-5	21-4	10-8	13-6	16-5	19-1
	Douglas fir-larch	#2	10-9	14-1	17-2	19-11	9-11	12-7	15-5	17-10
	Douglas fir-larch	#3	8-5	10-8	13-0	15-1	7-6	9-6	11-8	13-6
	Hem-fir	SS	10-9	14-2	18-0	21-11	10-9	14-2	18-0	21-11
	Hem-fir	#1	10-6	13-10	17-8	20-9	10-4	13-1	16-0	18-7
	Hem-fir	#2	10-0	13-2	16-10	19-8	9-10	12-5	15-2	17-7
	Hem-fir	#3	8-5	10-8	13-0	15-1	7-6	9-6	11-8	13-6
	Southern pine	SS	11-2	14-8	18-9	22-10	11-2	14-8	18-9	22-10
	Southern pine	#1	10-11	14-5	18-5	22-5	10-11	14-5	17-11	21-4
	Southern pine	#2	10-9	14-2	18-0	21-1	10-5	13-6	16-1	18-10
	Southern pine	#3	9-0	11-6	13-7	16-2	8-1	10-3	12-2	14-6
	Spruce-pine-fir	SS	10-6	13-10	17-8	21-6	10-6	13-10	17-8	21-4
	Spruce-pine-fir	#1	10-3	13-6	17-2	19-11	9-11	12-7	15-5	17-10
	Spruce-pine-fir	#2	10-3	13-6	17-2	19-11	9-11	12-7	15-5	17-10
	Spruce-pine-fir	#3	8-5	10-8	13-0	15-1	7-6	9-6	11-8	13-6
19.2	Douglas fir-larch	SS	10-8	14-1	18-0	21-10	10-8	14-1	18-0	21-0
	Douglas fir-larch	#1	10-4	13-7	16-9	19-6	9-8	12-4	15-0	17-5
	Douglas fir-larch	#2	10-1	12-10	15-8	18-3	9-1	11-6	14-1	16-3
	Douglas fir-larch	#3	7-8	9-9	11-10	13-9	6-10	8-8	10-7	12-4
	Hem-fir	SS	10-1	13-4	17-0	20-8	10-1	13-4	17-0	20-7
	Hem-fir	#1	9-10	13-0	16-4	19-0	9-6	12-0	14-8	17-0
	Hem-fir	#2	9-5	12-5	15-6	17-1	8-11	11-4	13-10	16-1
	Hem-fir	#3	7-8	9-9	11-10	13-9	6-10	8-8	10-7	12-4
	Southern pine	SS	10-6	13-10	17-8	21-6	10-6	13-10	17-8	21-6
	Southern pine	#1	10-4	13-7	17-4	21-1	10-4	13-7	16-4	19-6
	Southern pine	#2	10-1	13-4	16-5	19-3	9-6	12-4	14-8	17-2
	Southern pine	#3	8-3	10-6	12-5	14-9	7-4	9-5	11-1	13-2
	Spruce-pine-fir	SS	9-10	13-0	16-7	20-2	9-10	13-0	16-7	19-6
	Spruce-pine-fir	#1	9-8	12-9	15-8	18-3	9-1	11-6	14-1	16-3
	Spruce-pine-fir	#2	9-8	12-9	15-8	18-3	9-1	11-6	14-1	16-3
	Spruce-pine-fir	#3	7-8	9-9	11-10	13-9	6-10	8-8	10-7	12-4
24	Douglas fir-larch	SS	9-11	13-1	16-8	20-3	9-11	13-1	16-2	18-9
	Douglas fir-larch	#1	9-7	12-4	15-0	17-5	8-8	11-0	13-5	15-7
	Douglas fir-larch	#2	9-1	11-6	14-1	16-3	8-1	10-3	12-7	14-7
	Douglas fir-larch	#3	6-10	8-8	10-7	12-4	6-2	7-9	9-6	11-0
	Hem-fir	SS	9-4	12-4	15-9	19-2	9-4	12-4	15-9	18-5
	Hem-fir	#1	9-2	12-0	14-8	17-0	8-6	10-9	13-1	15-2
	Hem-fir	#2	8-9	11-4	13-10	16-1	8-0	10-2	12-5	14-4
	Hem-fir	#3	6-10	8-8	10-7	12-4	6-2	7-9	9-6	11-0
	Southern pine	SS	9-9	12-10	16-5	19-11	9-9	12-10	16-5	19-11
	Southern pine	#1	9-7	12-7	16-1	19-6	9-7	12-4	14-7	17-5
	Southern pine	#2	9-4	12-4	14-8	17-2	8-6	11-0	13-1	15-5
	Southern pine	#3	7-4	9-5	11-1	13-2	6-7	8-5	9-11	11-10
	Spruce-pine-fir	SS	9-2	12-1	15-5	18-9	9-2	12-1	15-0	17-5
	Spruce-pine-fir	#1	8-11	11-6	14-1	16-3	8-1	10-3	12-7	14-7
	Spruce-pine-fir	#2	8-11	11-6	14-1	16-3	8-1	10-3	12-7	14-7
	Spruce-pine-fir	#3	6-10	8-8	10-7	12-4	6-2	7-9	9-6	11-0

For SI: 1 inch = 25.4 mm, 1 foot = 304.8 mm, 1 pound per square foot = 0.0479 kPa.

Note: Check sources for availability of lumber in lengths greater than 20 feet.

a. Dead load limits for townhouses in Seismic Design Category C and all structures in Seismic Design Categories D_0, D_1 and D_2 shall be determined in accordance with Section R301.2.2.2.1.

GREEN OPTIONS

Chapter 5 is a great example of an IRC chapter that allows for green options within the actual prescriptive code requirements, even though they are not identified as green options per se. In addition to the prescriptive code criteria provided for traditional lumber, the IRC also provides a number of options that utilize alternative wood products. These products belong to a family of construction materials known as **engineered wood products** (manufactured wood products that have been tested and evaluated for acceptance under the code). Engineered wood products include construction materials such as end-jointed lumber (EJL), wood floor trusses, and prefabricated wood I-joists (see Figure 7-4).

The IRC also provides prescriptive code criteria for light-gauge steel floor framing. To its credit, the IRC has contained provisions for steel framing since its inception, beginning with the 2000 edition of the IRC. Steel floor framing offers a way to significantly reduce the wood needed to construct a dwelling for obvious reasons. Material standards, construction details, and span tables for steel framing are provided in the body of the code.

Regardless of the choice in floor framing materials, floors must be structurally sound and capable of resisting all required design loads. All of the alternative wood products explored in the following sections have been extensively tested and many have been evaluated to ensure that they will meet applicable code requirements and standards. The prescriptive code provisions for steel floor framing are also based on sound engineering principles and empirical data, making them a safe and acceptable way to meet code criteria.

Figure 7-4

Open-web wood floor trusses installed in a dwelling. Engineered floor trusses achieve long spans with relatively light weight.

COURTESY OF DOMINION TRUSS

Thus, a relatively practical way exists in the IRC to incorporate green construction into the floor framing without having to venture too far outside the basic code requirements: Simply utilize one of the "green" alternatives provided in the code. In doing so, the need for expensive engineering can be partially or, in some cases, completely eliminated because it has already been done by the manufacturers of the various engineered wood systems and those whose research culminated in the provisions for prescriptive steel framing.

WOOD FRAME FLOORS

It will come as no surprise that wood frame floors are made of wood. IRC Chapter 5 devotes a significant portion of the text to the use of solid-sawn wood framing, which is used extensively in most areas of the United States and Canada. Wood floor framing in use today encompasses much more than traditional solid-sawn lumber, however.

Wood floor framing also includes a variety of engineered lumber products that have gained popularity in recent years. In many of these cases, the IRC makes an outright allowance for these alternative wood products. Some engineered wood products use considerably less material than their solid-sawn counterparts. Other engineered wood products even make use of wood materials that would otherwise have been scrap or waste not too many years ago! Thus, they make excellent candidates for consideration when determining how to construct floors in the green dwelling project.

Section R502.1.3 End-Jointed Lumber

The IRC allows approved end-jointed lumber to be used interchangeably with solid-sawn lumber in floor framing. Also called finger-jointed lumber, EJL is characterized by fingerlike grooves between short segments of wood material that are glued together to form longer pieces (see Figure 7-5). End-jointed lumber is an engineered wood product. The specific manners in which the components are prepared and glued are prescribed by various standards. Additionally, the glues used in the fabrication of the joints are also required to meet certain standards as well. The U.S. Department of Commerce (DOC), through its Voluntary Products Standards program, publishes DOC PS 20-05, American Softwood Lumber Standard, which may prove useful, as it provides additional information regarding lumber products including grading and identification requirements.

Figure 7-5

End-jointed lumber to be glued and made into an engineered wood component.

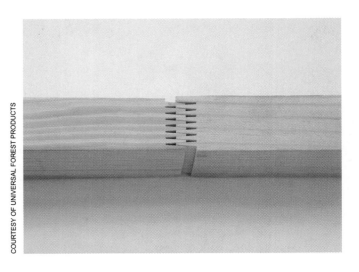

End-jointed lumber has a number of attractive properties. It is typically straighter than sawn lumber and contains fewer knots and less crook, bow, and twist when compared to its traditional cousin. Straighter lumber can mean less waste. EJL is available in a wide variety of wood species and traditional sizes including 2 inch by 4 inch (51 mm by 102 mm), 2 inch by 6 inch (51 mm by 152 mm), 2 inch by 8 inch (51 mm by 203 mm), 2 inch by 10 inch (51 mm by 254 mm), and 2 inch by 12 inch (51 mm by 305 mm) nominal dimensions. Because EJL is visually graded like traditional sawn lumber, it can be directly substituted for the traditional wood product of comparable species and grade. Because it is a manufactured product, it can also be ordered up to 40 feet (12,192 mm) in length.

Because the IRC expressly allows EJL material to be used as a substitute for solid-sawn lumber of the same species and grade, it is relatively easy to incorporate the material into the green construction project. This interchangeability means that the prescriptive span tables provided in the code for conventional lumber can be used without modification. Thus, the same span table that works for a traditional sawn 2 inch by 10 inch (51 mm by 254 mm) wood joist will work for an EJL 2 inch by 10 inch (51 mm by 254 mm) of the same species and grade.

Intent

The intent of this IRC section is to specifically allow the use of end-jointed lumber as a direct substitute for sawn lumber. The code further intends that, when substituting EJL for sawn lumber, that consideration be given to the species and grade of the material.

Green Decisions and Limitations

Shall I incorporate end-jointed lumber into the floor framing for my green dwelling?

EJL Increased Span Option

Because EJL is an engineered material, some manufacturers test their products to determine their performance and safe spans under various loading conditions. Where evaluation reports are available for EJL, increased spans may be allowed beyond those provided in the prescriptive provision of the IRC. Be sure to check these criteria closely when determining spans. EJL may actually perform better than sawn lumber in some cases.

Why It's Green

End-jointed lumber is green because it is made from short segments of lumber (scrap) that are fabricated into longer lengths. These short segments were, at one time, considered waste products, but are now manufactured into a useful construction material.

Section R502.1.5 Prefabricated Wood I-Joists

Although prefabricated wood I-joists ("I-joists" hereafter) have been around for approximately the last 40 years, it was not until the 1990s that they gained widespread acceptance for use in dwellings. These framing components are called I-joists because, in cross section, they resemble the letter "I," with the vertical

Figure 7-6

Prefabricated wood I-joists in various sizes. Note the "I" shape that is characteristic of these types of framing members.

COURTESY OF LP CORPORATION

center component being referred to as the "web" and the top and bottom horizontal components being referred to as "flanges" (see Figure 7-6). The shape of these wood framing members also resembles steel I-beams.

I-joists contain less wood material per foot of length than traditional lumber. This makes I-joists relatively lightweight compared to sawn lumber, making them somewhat easier to work and handle. I-joists are required to conform to ASTM D5055. Most major manufacturers of wood I-joists have done extensive testing on their products and most have obtained evaluation reports that provide the criteria for their use. These evaluation reports provide span tables for given loading conditions (similar to those provided in the IRC for sawn lumber) as well as conditions of use, limitations, and connection details.

I-joists can be used in virtually every floor framing condition as sawn lumber; however, wood I-joists are not manufactured to be installed in applications where they will be exposed to the weather or in other adverse conditions. Thus, they would not be a suitable choice for, say, an uncovered exterior deck in which the framing members would be directly subjected to the elements. For other unusual loading or environmental conditions, it is highly advisable to talk with the manufacturer about the specific conditions in which the wood I-joists will be used.

Intent

The intent of this IRC section is to specifically allow the use of prefabricated wood I-joists as an alternative to sawn lumber. The code further intends that, when substituting I-joists for sawn lumber, the specific requirements, conditions, and limitations established in the product evaluation reports must be considered.

TAKE NOTE!

Section R502.8.2 of the IRC prohibits cuts, notches, and holes bored into I-joists unless they conform strictly to the manufacturer's recommendations or where an engineering analysis has been provided to prove that the alteration to the I-joist will not cause an adverse effect. Manufacturer's recommendations for cuts, notches, and bored holes are typically found in the installation instructions or the manufacturer's specifications. Be sure to check these requirements closely before altering any engineered wood product.

Green Decisions and Limitations

Are prefabricated wood I-joists right for the floor framing in my green dwelling project?

How does the product evaluation report impact this decision?

Increased I-Joist Depth Option

Prefabricated wood I-joists are available in a variety of joist depths. Consider using joists that are deeper than traditional sawn lumber sizes as increased joist depths generally mean longer unsupported spans which, in turn, may mean that less additional lumber is needed for intermediate supports such as beams and girders (see Figure 7-7). Deeper joists also allow for more uncompressed insulation to be installed within the joist cavities. Compressed batt insulation performs at significantly lower levels than uncompressed batts. So, deeper joists not only reduce the amount of wood necessary for the project but also allow for improved insulation performance.

Why It's Green

Prefabricated wood I-joists are green because they use considerably less wood material than their solid-sawn cousins. The web material for these products is made from wood fiber that is derived from material that might otherwise be regarded as scrap or waste material. I-joists are lighter in weight for a given length of solid-sawn material, which can potentially reduce transportation costs and are usually straighter than sawn lumber, which means potentially less waste.

Figure 7-7

Wood I-joists shown in a typical floor framing configuration. In this case, they are framed into the side of a beam used for intermediate support.

COURTESY OF APA—THE ENGINEERED WOOD ASSOCIATION

STEEL-FRAMED FLOORS

Floors framed out of light-gauge, cold formed steel are strong and durable. Although not widely used in residential construction, steel framing is an option that is gaining popularity of late. Steel framing used in floors is similar in layout to that of a wood floor, in that steel joists are spaced at 12-inch (305-mm), 16-inch (406-mm), and 24-inch (610-mm) intervals. Like their wood counterparts, the allowable spans for steel framing members are also a function of the depth of the joist, the particular loading conditions and the grade of the material used, and the gauge of the steel used in the fabrication of the joist. In this case, however, the "grade" is determined by the structural properties inherent in the steel used in the manufacture of the joist as opposed to the species of a tree as is the case with wood.

Steel floor framing components must be properly identified with a mark similar to that used in wood framing. Such markings are required to identify the manufacturer of the steel joist, the uncoated steel thickness, the minimum coating designation (for corrosion resistance), and the minimum yield strength, in kips per square inch (Mpa). A **kip** is a unit of measure used in engineering equal to a kilopound or 1,000 pounds.

It is here, however, where the similarities end. Steel floor framing requires special tools to cut, bore, and otherwise work the components. It is typically fastened together using self-tapping, self-drilling screws. Traditional nails are not an appropriate means of connection for steel—even if they could be driven through the metal with a hammer! Cutting is typically done with a steel nipper, with saws equipped with special blades or other hydraulic tools with hardened blades and bits suitable for cutting steel.

Other special considerations are required for steel floor framing as well. For example, where steel floor joists are supported by steel frame walls, each floor joist must be aligned directly over the supporting wall stud below. The maximum tolerance for in-line placement is 0.75 inch (19 mm) per Section R505.1.2 of the IRC.

Section R505.1 Cold Formed Steel Floor Framing

This IRC section charges all cold formed steel floor framing components to comply with the provisions of Section R505. Although not explicitly stated, the section authorizes steel framing to be used in floor framing applications. Section R505.1.1, Applicability Limits, introduces a restriction, however, on the use of steel framing under the prescriptive provisions.

This section restricts the use of steel framing to dwellings that are not more than two stories in height; not greater than 60 feet (18,288 mm) in length measured perpendicular to the span of the joist; and not greater than 40 feet (12,192 mm) in width measured parallel to the joist span. The IRC further restricts the use of steel framing in floors to sites with maximum wind speeds of 110 mph (49.2 m/s), exposure A, B, or C, and snow loads not greater than 70 pounds per square foot (3.35 kPa). Wind exposures are rated under the provisions of the code by letter designation, where "A" designates areas with physical obstructions and topography that are likely to reduce wind speeds (least affected by wind) and "D" designates open, flat areas where wind speeds are likely to increase (most affected by wind).

Does this mean that steel framing is not allowed in locations where these limits are exceeded? No; what it means is that the prescriptive requirements established in the IRC cannot be used in those instances. A structure can always be engineered in lieu of the prescriptive requirements in the code, however.

Steel load bearing components are generally of two different configurations: C section and track section. The C section shape receives its name from the shape of the joists which resembles the letter "C." Track section is similar in appearance to a C section but lacks the rolled "lip" as seen in Figure 7-8a and 7-8b. Steel C-section joists are used in a variety of configurations that are constructed to resist the loads encountered in the typical dwelling.

In some cases, C sections are used back to back and fastened together in the web to form an "I" shape, similar to that described in the preceding section. In other cases, C sections are combined with track sections to form built-up headers and girders. Again, these components are fastened together to form strong, lightweight structural supports for floor components.

The IRC provides several useful construction details that describe how to assemble various components into headers and trimmer joists (see Figure 7-9). These details are meant to be prescriptive in nature, which means that engineering is not required where the particular use of the steel framing is within the limits established in the code.

Figure 7-8a, b
Figures R505.2(1)
C-shaped section and
R505.2(2) track section.

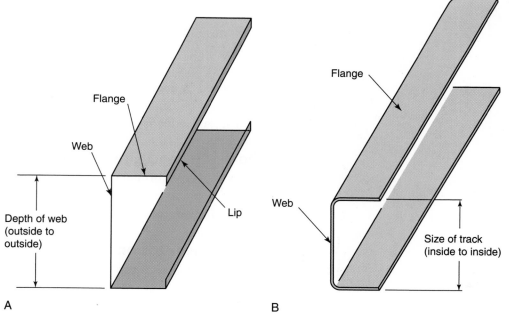

A   B

COPYRIGHT 2009 IRC®

Figure 7-9

Figure R505.3.8(3) cold formed steel floor construction: floor header to trimmer connection— 6-foot opening.

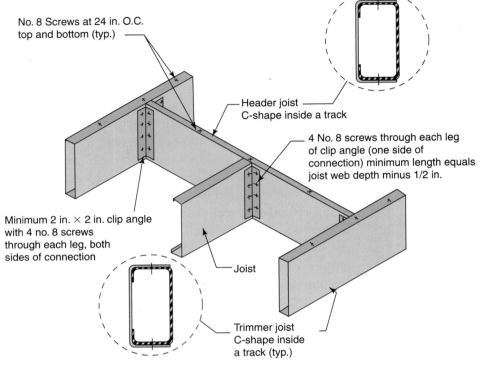

No. 8 Screws at 24 in. O.C. top and bottom (typ.)

Header joist C-shape inside a track

4 No. 8 screws through each leg of clip angle (one side of connection) minimum length equals joist web depth minus 1/2 in.

Minimum 2 in. × 2 in. clip angle with 4 no. 8 screws through each leg, both sides of connection

Joist

Trimmer joist C-shape inside a track (typ.)

For SI: 1 inch = 25.4mm

Why It's Green

Cold formed steel floor framing significantly reduces the amount of wood products needed to construct a code-compliant floor system, which potentially lessens impacts to forests. Steel framing is fastened with screws that easily accommodate deconstruction when the building has served its useful life. Steel is easily recycled.

Intent

The intent of these IRC provisions is to provide prescriptive requirements for steel floor framing. The code further intends that, where cold formed steel framing is used, it is used within the applicability limitation imposed by Section R505.1.1. The code further intends that steel floor framing be capable of accommodating all imposed loads and distributing those loads to the supporting structural elements.

Green Decisions and Limitations

Shall I incorporate cold formed steel framing into my green dwelling project?
How do the requirements of the IRC affect this decision?

Steel Floor Truss Option

Open-web steel floor trusses can easily be incorporated into the floor design. Steel floor trusses are required to meet the American Iron and Steel Institute (AISI) Standard for Cold-Formed Steel Framing—Truss Design (COFS/Truss). Steel trusses can span considerable distances without intermediate supports and the open-web configurations

Figure 7-10

A type of open-web steel truss used in the floor framing system of a dwelling. Note the large, open spaces that allow for easy routing of ducts, pipes, and wires.

COURTESY DELMAR/CENGAGE LEARNING

make routing ducts, pipes, and wire a relatively easy operation (see Figure 7-10). They share the advantages that other steel framing components have as well. Some steel floor trusses use wood for the top and bottom flange, making the installation of plywood for the floor and gypsum board for the ceiling a simple operation.

CHAPTER 7 APPLIED TO THE SAMPLE PROJECT

Chapter 7 pertains to floors and floor framing systems of sawn wood, steel, and engineered wood systems. Additionally, Chapter 7 addresses concrete floors (on ground). The following analysis will demonstrate the application of some IRC code provisions for floors with consideration given for the proposed scope of work and our design philosophy.

Site Considerations

The flat site makes the use of a concrete slab well suited to the task of providing a floor for the sample project. It will be necessary to remove all vegetation, top soil, and foreign matter from the slab area within the foundation walls per Section R506.2. Where groundwater is a problem, special drainage may be required. The local jurisdiction may be aware of areas where groundwater and other site hazards may be present, so it is wise to add this to our list of discussion points.

Building Considerations

Rammed-earth walls are massive and their extraordinary weight requires special consideration in the structural design. The design professional will consider the footing design and the best way to tie the slab floor to the footings, resulting in either a **monolithic foundation** (footings, foundations, and slab poured as one unit) or a slab that is completely separate from the footings. Where fill is necessary under the slab floor area, it must be properly compacted to provide adequate support for the slab. A vapor retarder may be required under the slab to prevent moisture vapor from entering the dwelling. Reinforcement for the slab may be required as well. In this case, the sample project is a single-story dwelling, so no second-floor framing is required.

CHAPTER SUMMARY

Chapter 5 of the IRC addresses the various code requirements for floor framing systems. It provides standards for materials and methods of installation for both wood and steel floor construction materials and information on nonstructural slab floors and under-floor spaces. Many of the engineered wood products covered in the chapter offer green alternatives. Following are highlights of the chapter.

- Chapter 5 of the IRC provides prescriptive code requirements for both wood and steel frame floors.
- Criteria for floor framing systems are provided in text and tabulated form.
- Span tables are provided for both wood and steel construction.
- Wood and steel floor framing systems collect and distribute loads to supporting elements.
- A wood or steel floor joist's allowable span is a function of several criteria including required load, spacing of the member, the species and grade of the material if wood or the grade of the steel, and allowable deflection criteria.
- Many of the allowed applications available for floor framing are already "green" alternatives such as engineered products and can simply be chosen as an allowed option.
- Manufacturer's installation guidelines for engineered wood products must be observed.

WALLS

This chapter corresponds to IRC Chapter 6, Wall Construction

learning objective

To know and understand how walls are regulated under the provisions of the IRC and to explore options for green wall construction.

IRC CHAPTER 6—OVERVIEW

Chapter 6 of the International Residential Code regulates the construction of walls for dwellings. Walls enclose the living space where we live our lives. They provide security from intrusion, privacy when needed, and important protection from the elements. Moreover, walls form an important part of the structural system that supports a dwelling. Walls collect loads from floors and roofs and distribute these forces into the supporting elements. The IRC establishes code criteria for walls to ensure that loads are safely and effectively transferred to floors or other resisting elements such as the foundation.

Walls are made up of vertical, structural members called studs (see Figure 8-1). The ends of studs are typically capped with wood plates of the same size and material as the stud itself. The top plate is typically doubled and the bottom plate, often called a sole plate, is usually a single member. The IRC now also allows optimized or advanced framing techniques such as single top plates and two-stud corners. These framing techniques reduce the amount of wood necessary to frame the wall, which results in increased insulation and improved energy efficiency. In either case, the top plates in load-bearing walls create a bearing surface on which the floor framing and roof framing members rest. Nonload-bearing walls serve as partitions only and do not carry or distribute loads from or to other parts of the structure.

Like floors, load-bearing wall studs support or resist two primary types of load: gravity loads and lateral forces. Gravity loads act on walls at all times and cause these vertical framing members to compress under the applied force. As the studs

Figure 8-1

Studs forming an exterior bearing wall. Note the single bottom plate and the doubled studs which, in this case, support the end of a beam.

compress axially (i.e., through the long axis of the stud), stress begins to develop within the wood fibers to resist the weight of people, appliances, and even the piano, as well as dead loads from the structure itself.

The structural properties for the particular grade and species of lumber come into play here, as well as stud size, and determine in part how much load the stud can withstand. The amount of weight the stud can truly support, however, is also a function of how tall the stud is in relationship to its **unbraced length** (the length of the stud between supports). To understand this phenomenon, imagine that you have a yardstick in your hand and that it is standing on its end, like a stud would be when framed in a wall. Now, also imagine that you try to support your weight by pressing down on the yardstick (the unbraced length in this case is the distance between your hand and the floor). As you will soon discover, the yardstick will buckle long before the other end of it crushes under your weight. Tall, skinny studs behave the same way as the yardstick and tend to buckle under a load before the wood fiber crushes. Fortunately, wood structural sheathing like plywood oriented strand board (OSB) continuously braces wall studs, allowing them to support great weights without difficulty.

Wall studs also resist lateral forces that come from wind and seismic loads through load transfer from other structural systems. Recall from Chapter 1 of this text that lateral forces are intermittent; that is, they are not always present. However, when the wind blows hard or where an earthquake occurs, the forces generated can be staggeringly high. Lateral forces cause the studs to bend and flex in response to the applied force.

Again, structural sheathing helps tremendously here, as the plywood stiffens the wall and resists the forces received from the applied wind and seismic loads. Studs in walls must resist not only the ever-present gravity loads but also the lateral forces when they occur. In other words, the wall construction must simultaneously resist both vertical and lateral loads. These combined loading effects must be considered in the design of a wall, as well.

Stud heights, sizes, and spacing for prescriptive wall framing are determined by the provisions of IRC Table R602.3(5) of the code (see Figure 8-2). Requirements and limitations for studs are dependent on whether the studs are bearing or nonbearing. It stands to reason that load-bearing studs must be more closely spaced than their nonload-bearing counterparts. For example, note that a 2 inch by 4 inch (51 mm by 102 mm) load-bearing stud spaced at 24 inches (610 mm) on center, supporting

Figure 8-2

Table R602.3(5) Size, height and spacing of wood studs.

TABLE R602.3(5)
SIZE, HEIGHT AND SPACING OF WOOD STUDS[a]

STUD SIZE (inches)	BEARING WALLS					NONBEARING WALLS	
	Laterally unsupported stud height[a] (feet)	Maximum spacing when supporting a roof-ceiling assembly or a habitable attic assembly only (inches)	Maximum spacing when supporting one floor, plus a roof-ceiling assembly or a habitable attic assembly (inches)	Maximum spacing when supporting two floors, plus a roof-ceiling assembly or a habitable attic assembly (inches)	Maximum spacing when supporting one floor height[a] (feet)	Laterally unsupported stud height[a] (feet)	Maximum spacing (inches)
2 × 3[b]	—	—	—	—	—	10	16
2 × 4	10	24[c]	16[c]	—	24	14	24
2 × 4	10	24	24	16	24	14	24
2 × 5	10	24	24	—	24	16	24
2 × 6	10	24	24	16	24	20	24

For SI: 1 inch = 25.4 mm, 1 foot = 304.8 mm, 1 square foot = 0.093 m².
a. Listed heights are distances between points of lateral support placed perpendicular to the plane of the wall. Increases in unsupported height are permitted where justified by analysis.
b. Shall not be used in exterior walls.
c. A habitable attic assembly supported by 2 × 4 studs is limited to a roof span of 32 feet. Where the roof span exceeds 32 feet, the wall studs shall be increased to 2 × 6 or the studs shall be designed in accordance with accepted engineering practice.

a roof and ceiling only, is limited to 10 feet (3,048 mm) in height, whereas a 2 inch by 4 inch (51 mm by 102 mm) nonload-bearing stud at the same spacing can be 14 feet (4,267 mm) in height.

Stud heights and spacings are also a function of the particular species and grade of the lumber material. Some species of lumber are stronger than others and are thus better suited for certain structural applications. In the western United States, Douglas fir is commonly used for studs in construction. In other areas of the country, southern yellow pine is used. Largely, the type of lumber used in construction is a matter of what is available regionally. Table R602.3(5) is based on studs with a minimum assumed grade of No. 3, standard or stud grade lumber, and with the assumption that the species will be used in prescriptive construction, with bearing walls not taller than 10 feet (3,048 mm) in height. Studs of a higher grade may, by engineering analysis, be allowed at heights greater than that allowed in the table.

Fortunately, the IRC makes the determination of proper stud height, size, and spacing easy through the use of tables such as the one in Figure 8-2. These tables have been developed through years of gathering empirical data and through successful use in literally hundreds of thousands of successful construction projects. Without such a table, an engineering analysis would be required for each construction project based on actual loading conditions.

GREEN OPTIONS

Like the previous one, Chapter 6 is another great example of an IRC chapter that allows for green options within the body of the text. In addition to the prescriptive code criteria for walls built out of traditional lumber, the IRC also provides a number of choices that use alternative construction materials, including masonry, structural insulated panels (SIPs), and products known as **insulating concrete forms (ICF)** (concrete forms made of rigid foam plastic, left in placed once filled). ICF products include construction materials and options such as flat ICF, waffle-grid ICF, and screen-grid ICF (see Figure 8-3).

As we saw with floors, the IRC also provides prescriptive code criteria for light-gauge steel wall framing. Steel wall framing can significantly reduce the wood needed to construct a dwelling although not entirely, as the IRC assumes that steel frame walls will be sheathed with plywood, OSB, or particle board approved for structural applications (see Figure 8-4). Material standards, construction details, wall support tables, and header span tables for steel framing are provided within the body of the code.

As with conventional wood wall framing materials, walls built from alternative materials must be structurally sound and capable of receiving and resisting all required design loads. All of the alternative construction materials explored in the following sections have been extensively tested and evaluated to ensure that they will meet all applicable code requirements and standards. The prescriptive code provisions for steel wall framing are also based on sound engineering principles and empirical data as they are for floors, making them a safe and acceptable way to meet code criteria.

Figure 8-3

A dwelling in which the first- and second-story walls are constructed out of insulating concrete forms (ICF).

COURTESY OF GREENBLOCK WORLDWIDE CORP

Figure 8-4

Steel studs used to frame the gable-end wall of a dwelling. Note that plywood is used as sheathing and fastened to the studs with screws.

Because the IRC provides several prescriptive options for green wall construction, it is relatively easy to incorporate green methods into the wall system without having to venture too far outside the basic code requirements: Simply choose one of the green options provided in the code. In doing so, the need for expensive engineering can be partially or, in some cases, completely eliminated because it has either already been done by the manufacturers of the various alternative wall framing products or by those whose research culminated in the provisions for the prescriptive framing options found in the code.

WOOD FRAME WALLS

For the vast majority of conventional construction that occurs in the United States and Canada, wood is the material of choice for both interior and exterior walls. IRC Chapter 6 devotes a significant portion of the text to the use of solid-sawn wood framing which is used extensively in most areas of the United States. Traditional, solid-sawn wood is not the only choice for wood wall construction, however.

As we saw with floors, wood wall framing also includes a variety of engineered lumber products that have gained popularity in recent years. In many of these cases, the IRC makes an outright allowance for these alternative wood products, such as with end-jointed studs. Other engineered wood products utilize wood materials made from wood by-products and waste materials from other manufacturing processes. There is even an application for wood I-joists to be used vertically, as a replacement for conventional wood studs. Thus, alternative wood products make excellent candidates for consideration when planning the wall construction methods in the green dwelling project.

Section R602.1.1 End-Jointed Lumber

The IRC allows approved end-jointed lumber (EJL) to be used interchangeably with solid-sawn lumber in wall framing. End-jointed lumber studs can be bored, cut, and notched just like their solid-sawn cousins. They can also be doubled for use as posts or used as blocking and trimmers for the support of headers. End-jointed lumber studs can be worked with traditional tools the same as ordinary studs. In short, they can be used in any application where a standard wood stud can be used for the same species and grade of material (see Figure 8-5).

End-jointed lumber has a number of attractive properties for use in walls, just as it does in floor framing applications. It is typically straighter than sawn lumber and contains less crook, bow, and twist when compared to traditional wood studs. End-jointed lumber is available in a wide variety of wood species and traditional sizes, including 2 inch by 4 inch (51 mm by 102 mm), 2 inch by 6 inch (51 mm by 152 mm), and 2 inch by 8 inch (51 mm by 203 mm) nominal dimensions. Because EJL is visually graded like sawn lumber, it can be directly substituted for the traditional wood studs used in prescriptive walls. The availability of long lengths makes it attractive for use in gable-end walls where balloon framing is preferred and in other applications where stud lengths exceed standard sizes.

Because the IRC expressly allows EJL material to be used as a substitute for solid-sawn lumber of the same species and grade, it is relatively easy to incorporate this material into the green construction project. This interchangeability means that the prescriptive wall height and stud spacing table provided in the code for conventional lumber can be used without modification. Additionally, the various header and girder span tables will work without modification, as well. Thus, the same stud height and spacing that works for a traditional 2 inch by 6 inch (51 mm by 152 mm) wood stud will work for an EJL 2 inch by 6 inch (51 mm by 152 mm) of the same species and grade.

Intent

The intent of this IRC section is to specifically allow the use of end-jointed lumber as a direct substitute for sawn lumber. The code further intends that, when substituting EJL for sawn lumber, that consideration be given to the species and grade of the material.

Figure 8-5

Finger-jointed lumber shown in both a glued and unglued application. Note that the joint resembles interwoven fingers, hence the name.

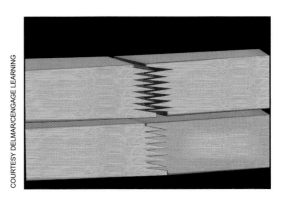

COURTESY DELMAR/CENGAGE LEARNING

Green Decisions and Limitations

Shall I incorporate end-jointed lumber into the wall framing for my green dwelling?

How do the requirements of the IRC affect this decision?

EJL Increased Span Option

Some EJL studs have superior grade characteristics because the studs are engineered. Obvious defects such as knots and other grain defects can be avoided as the studs are manufactured, because the segments are visually selected for use. Wood segments with major flaws are simply not used. Better grades equate to improved structural properties which typically mean longer allowable spans under given loading conditions. This can be useful in a variety of applications. Where longer stud lengths are desirable, such as with gable-end walls and for use in interiors with vaulted ceilings, longer studs may mean that walls can be balloon framed rather than platform framed, which can potentially reduce labor. Be sure to check grade criteria closely when determining spans.

OTHER GREEN OPTIONS FOR WOOD WALLS

Some alternative wood products are not specifically addressed under the wall framing provisions in the IRC but are certainly worth exploration. For example, SIPs are used in many areas of the United States. Recall from a prior chapter that SIPs are essentially a sandwich panel with a thick layer of rigid foam insulation between two layers of OSB or other structural sheathing. Alternatives also include timber frame and other methods.

Laminated Veneer Lumber Used as Studs

Laminated veneer lumber (LVL), or composite lumber hereafter, is typically used as horizontal framing in floors, roofs, and ceilings. A lesser-known but equally practical use for LVL is in vertical framing applications such as studs (see Figure 8-6).

LVL contains less solid wood material per foot of length than traditional lumber, but has inherent structural properties that can greatly exceed ordinary wood studs of comparable size. This makes LVL an excellent choice for use in extremely tall walls where ordinary wood framing is not practical. Composite lumber is an engineered wood product and must comply with the manufacturers' listings.

Most major manufacturers of composite wood products for wall applications have done extensive testing on their products and most have obtained evaluation reports that provide the criteria for their use in walls. These evaluation reports provide height and spacing criteria in tables for given loading conditions (similar to those provided in the IRC for sawn lumber) as well as conditions of use, limitations, and connection details. These reports also identify specific conditions under which the products cannot be used.

LVL can be used in many wall framing conditions like sawn lumber; however, composite wood products are not manufactured to be installed in applications where they will be exposed to the weather or in other adverse conditions. Thus, they would not be a suitable choice for, say, columns or other vertical supports

Why It's Green

End-jointed lumber is green because it is made from short segments of lumber (typically scrap materials) that are fabricated into longer, useful lengths. Because EJL is typically straighter and truer than conventional wood studs, there can be potentially less material waste due to the wood twisting or warping and making it otherwise unusable.

Figure 8-6

Laminated veneer lumber (LVL) used in wall framing. Engineered wood products have gained popularity in recent years for their impressive structural properties and consistent quality.

COURTESY OF ILEVEL

that would be directly subjected to the elements. For other unusual loading or environmental conditions, it is highly advisable to talk with the manufacturer about the specific conditions to which the LVL will be exposed.

Green Decisions and Limitations

Is LVL right for the wall framing in my green dwelling project?

How does the product evaluation report impact this decision?

Why It's Green

LVL or composite lumber is green because it uses wood fiber and adhesives to make a wood product superior to sawn lumber. The use of wood fiber allows scrap material to be utilized instead of wasted. LVL is straighter and truer than sawn lumber, which means less waste due to natural defects in the wood.

Structural Insulated Panels

SIPs are a durable, highly efficient panelized form of construction. When used in wall applications, SIPs are stood on edge and installed on a wood plate that is fastened to a floor or directly to a foundation or slab (see Figure 8-7). At first glance, SIPs do not appear to be capable of supporting great weights (after all, they are largely made of rigid foam); however, they can support significant loads and are engineered to perform comparably to any standard framed wall built of traditional materials.

SIPs use special adhesives to bond the foam core to the outer layers of structural sheathing. This permanent bond enables the SIP to collect and distribute loads much like other forms of construction. Panels are designed with channels that run through the foam to allow electrical wiring and plumbing piping to be routed through the panels. SIPs are highly energy efficient and outperform standard wood frame walls in many cases.

Figure 8-7

Structural insulated panels (SIPs) shown in a typical roof-wall configuration. Note the foam core "sandwiched" between two sheets of oriented strand board (OSB) in each panel.

COURTESY OF AFM CORPORATION®

New to the 2009 edition of the International Residential Code are prescriptive code provisions for the use of SIPs in wall construction. The prescriptive provisions can be found in Section R613 and include information on standards for fabrication, wall height limitations, and assembly requirements. The incorporation of SIPs into the prescriptive body of the code is a major step forward in recognizing what were once AMM and recognizing the practicality and usefulness of SIPs in conventional construction.

Green Decisions and Limitations

Are SIPs right for the wall framing in my green dwelling project?

STEEL FRAME WALLS

Walls framed out of light-gauge, cold-formed steel perform just as well as wood stud walls. Although not widely used in residential construction, steel framing is an option that is gaining popularity as those interested in green building continue to look for alternatives to wood. Steel framing used in walls is similar in layout to that of a wood wall, in that steel studs are spaced at 16-inch (406-mm) and 24-inch (610-mm) intervals. Like their wood counterparts, the allowable heights and spacings for steel framing members are also a function of the depth of the joist, the particular loading conditions involved, and the grade of the steel used. As we saw with steel floors, the "grade" is determined by the particular type of steel used in the manufacture of the stud as opposed to the species of a tree.

Steel wall framing components must be properly identified with a label, stencil, or other mark similar to that used in wood framing. Such markings are required to identify the manufacturer of the steel joist, the uncoated steel thickness, the minimum coating designation (for corrosion resistance), and the minimum yield strength, in kips (kilopounds; 1 kip equals 1,000 pounds) per square inch.

Why It's Green

SIPs are green because they contain considerably less wood material than walls formed from solid-sawn wood framing. The foam core within the SIP is highly energy efficient. SIPs are available in most locations, reducing the need to transport them over long distances.

Figure 8-8

Figure R603.2.5.1 Web holes.

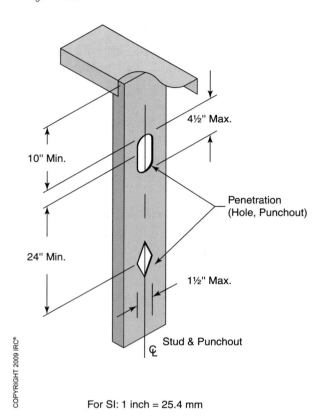

4½" Max.

10" Min.

Penetration
(Hole, Punchout)

24" Min.

1½" Max.

Stud & Punchout

℄

For SI: 1 inch = 25.4 mm

It is here, however, where the similarities with wood studs end. Steel wall framing requires special tools to cut, bore, and otherwise work the components. Flanges must not be cut as required by Section R603.3.4. Holes and cuts in the stud, which are often necessary for electrical wiring and plumbing piping, must be carefully placed in the web of the stud as seen in Figure R603.2.5.1. Web holes must occur through the centerline of steel stud (see Figure 8-8). Often, these holes are punched into the web during the manufacturing process, which ensures accurate placement and reduces the need for field modification.

Other special considerations are required for steel wall framing as well. For example, steel wall studs must directly align with floor joists, roof rafters, trusses, and other structural elements, requiring careful layout and placement during installation. Wall studs must also align with other wall studs in stories above and below the studs in question. The maximum tolerance for in-line placement is 0.75 inch per Section R505.1.2 of the IRC. This means that careful placement of steel framing members is a necessity when using steel to frame a residential structure.

Section R603.1 General Requirements

This IRC section charges all cold-formed steel wall framing components to comply with the provisions of Section R603 Steel Wall Framing. Although not explicitly stated, the section authorizes steel framing to be used in wall framing applications. Section R603.1.1, Applicability Limits, introduces the same restriction for steel wall studs as that imposed for floors when using steel framing under the prescriptive provisions of the IRC: 60 feet (18,288 mm) by 40 feet (12,192 mm); two stories in height; ground snow load 70 pounds (3.35 kPa) maximum; and wind speeds limited to 110 mph (49 m/s), exposure A, B, or C. Where these prescriptive limitations cannot be met, the IRC requires steel wall framing to be designed.

Steel load-bearing components are generally of two different configurations: C section and track section (see Chapter 7 of this text for a review). Steel C-section studs are used in a variety of configurations, constructed to resist the gravity and lateral loads encountered in the typical dwelling. In some cases, C sections are used in a back-to-back manner and fastened together in the web to form an I shape, similar to that described in the preceding section. In other cases, C sections are combined with track sections to form headers and columns. Again, these components are fastened together to form strong, lightweight structural supports for wall components.

Figure 8-9

Figure R603.6(1)
Box beam header.

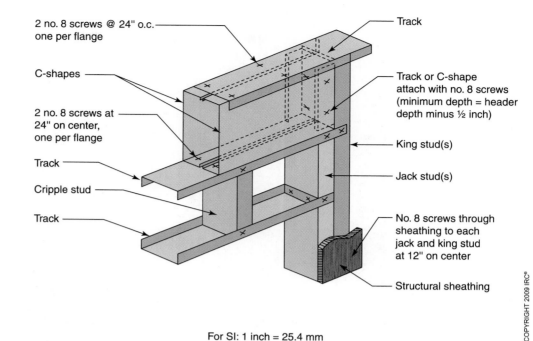

2 no. 8 screws @ 24" o.c. one per flange

C-shapes

2 no. 8 screws at 24" on center, one per flange

Track

Cripple stud

Track

Track

Track or C-shape attach with no. 8 screws (minimum depth = header depth minus ½ inch)

King stud(s)

Jack stud(s)

No. 8 screws through sheathing to each jack and king stud at 12" on center

Structural sheathing

COPYRIGHT 2009 IRC®

For SI: 1 inch = 25.4 mm

The IRC provides several useful construction details that describe how to assemble various components into headers (see Figure 8-9). These details are meant to be prescriptive in nature, which means that engineering is not required where the particular use of the steel framing is within the limits established in the code.

Intent

The intent of these IRC provisions is to provide prescriptive requirements for steel wall framing. The code further intends that where cold-formed steel framing is used, it is used within the applicability limitations imposed by Section R603.1.1. The code further intends that steel wall framing members be capable of accommodating all imposed loads and distributing those loads to the supporting structural elements.

Green Decisions and Limitations

Shall I incorporate cold-formed steel framing into my green dwelling project?

How do the requirements of the IRC affect this decision?

Why It's Green

Cold-formed steel wall framing significantly reduces the amount of wood products needed to construct a code-compliant wall system, which potentially lessens impacts to forests. Steel wall framing is fastened with screws that easily accommodate deconstruction when the building has served its useful life, allowing for the steel wall components to be easily recycled. Steel wall studs can be lighter than solid-sawn members of the same length, which can potentially reduce shipping costs.

Figure 8-10

Spray foam insulation being applied to an exterior steel stud wall. Note how the foam insulation expands to fill all crevices and voids in the wall cavity.

COURTESY OF MARK MECKLEM, MIRANDA HOMES

Increased Insulation Option

Steel wall studs take up less physical space in a wall than wood studs. This allows for more insulation to be added within a given wall cavity, especially when using spray-applied foam insulation (see Figure 8-10). Increased insulation generally improves the thermal performance of a dwelling by reducing heat, loss through the wall; however, steel is also an excellent conductor of heat, which can potentially offset the gains from the additional insulation. This problem is usually solved through the addition of a layer of foam sheathing on the exterior of the structure to protect the steel studs from direct exposure to extremes of temperature.

INSULATING CONCRETE FORMS

Insulating concrete forms have gained tremendous popularity in recent years and are now used throughout the United States in both foundations and exterior wall applications. An insulating concrete form, as the name implies, is a concrete form manufactured out of rigid foam, designed to be left in place after the concrete is poured.

These concrete forms come in three primary types: flat ICF, waffle-grid ICF, and screen-grid ICF. They are characterized by the configuration of the inner core, which varies in size and shape.

Flat ICF systems utilize a flat foam shape within the core and produce a wall that is identical to a standard concrete wall poured in the traditional way. Because the concrete wall produced with a flat ICF is of the same uniform thickness as a conventionally formed concrete wall, the code treats them identically. Flat ICF walls can be used in any application that plain, poured concrete walls can be used. Provisions for flat ICF are provided in Section R611.3.1 of the IRC.

Waffle-grid ICF differs from the flat ICF in that the core inside the ICF consists of a series of honeycomb-like voids that are reinforced with steel and filled with concrete. A spacer inside the form helps the form keep its shape as it is filled with concrete. The name "waffle grid" comes from the appearance of the concrete surface inside the form once poured, with thicker ribs of concrete and thinner flat concrete areas between the grids much like the surface of a breakfast waffle (see Figure 8-11). Provisions for waffle-grid ICF are provided in Section R611.3.2 of the IRC.

Screen-grid ICF differs from the previous two in that there are no flat concrete areas connecting the concrete ribs. With screen grid, there is an actual void between the reinforced concrete ribs. Provisions for screen-grid ICF are provided in Section R611.3.3 of the IRC.

Sections R611.3.1, R611.3.2, and R611.3.3 Flat, Waffle-Grid, and Screen-Grid ICF

Prescriptive requirements for ICF are provided in both text and tabular formats. ICF may be used prescriptively as outlined in the code or in accordance with their product evaluation reports. Most major manufacturers now offer ICF products that have been thoroughly tested and can be used in a variety of applications.

Like steel frame walls, ICF are subject to certain applicability limitations per Section R611.2. ICF can be used prescriptively. This section restricts the use of ICF to dwellings that are not more than two stories in height; not greater than 60 feet (18,288 mm) in length in any direction; floors with clear spans not greater than 32 feet (9,754 mm); and roofs with clear spans not greater than 40 feet (12,192 mm). Additionally, the IRC further restricts the use of ICF to sites with maximum wind speeds of 130 mph (58 m/s) for wind exposure B; 110 mph (49 m/s) for wind exposure C; and 100 mph (45 m/s) for wind exposure D. The use of ICF is further limited to Seismic Design Categories A or B for one- and two-family dwellings and townhouses and Category C for one- and two-family dwellings only.

Wall heights are provided in tables such as the one seen previously in Figure 8-2. The table establishes the maximum wall heights depending on the type of ICF

Figure 8-11

The typical concrete configuration for a waffle-grid ICF application (forms removed). The name is derived from the waffle-like appearance of the concrete concealed inside the forms.

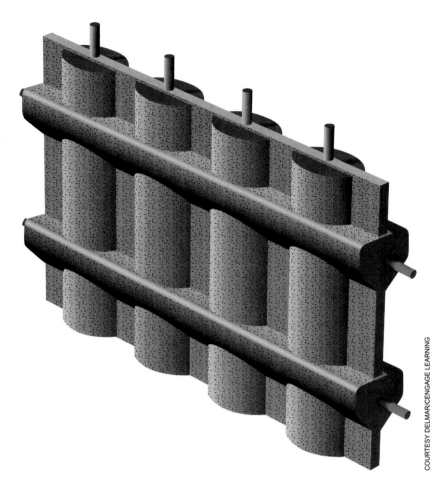

COURTESY DELMAR/CENGAGE LEARNING

used. The horizontal and vertical spacing of the cores within the ICF are defined as well.

Intent

The intent of these IRC sections is to provide minimum standards for construction materials and methods used in ICF construction. The code further intends that ICF conform to the applicability limitations outlined in Section R611.2. Additionally, the code intends that ICF be installed per the requirements of the manufacturer's listing and product approval reports.

Green Decisions and Limitations

Shall I incorporate ICF into the walls for my green dwelling project?
How do the requirements of the IRC affect this decision?

Exterior Wall Option

ICF are well insulated and make for an energy-efficient exterior envelope. ICF can be used in one- and two-story applications under the prescriptive provisions of the IRC. Concrete is an extremely durable construction material that is relatively maintenance free. Exterior and interior surfaces can be finished in traditional materials to have a conventional appearance. Some ICF use up to 30%

less concrete than other types and are worth considering for this aspect as well.

CHAPTER 8 APPLIED TO THE SAMPLE PROJECT

Chapter 8 pertains to wall systems of sawn wood, light-gauge steel, masonry, SIPs, and other systems. Additionally, Chapter 8 addresses ICF walls in prescriptive fashion. Our project requires an alternative materials and methods (AMM) proposal, however, due to the proposed use of stabilized, reinforced rammed earth for the wall material. The following analysis will demonstrate the application of some of the IRC provisions for walls with consideration given for the proposed scope of work and our design philosophy.

Why It's Green

ICF used in the construction of walls is a wood-free alternative. ICF produce a wall made of concrete that is very strong and durable. ICF are available throughout the United States, which means they are available locally, reducing the environmental effects of shipping building products over long distances. ICF can produce a highly energy-efficient wall.

TAKE NOTE!

ICF is covered with foam plastic that must be covered with an approved flame spread and thermal barrier such as gypsum wallboard in occupied areas. These barriers slow the spread of fire and provide a layer of thermal protection between the foam plastic and the room. In this case, the thermal barrier is not an added measure of energy conservation; rather, it is a fire safety measure required by the code. Thus, in areas such as unfinished basements it may be necessary to cover the foam with gypsum board or other approved materials to meet the requirements of Sections R316.3 Surface-Burning Characteristics and R316.4 Thermal Barriers.

Site Considerations

Recall the project scope which indicated that, due to site characteristics, it will be necessary to locate one of the exterior walls of the sample project 3 feet (914 mm) from a property line. As such, 1-hour-rated fire-resistive construction will be required to satisfy the requirements of Section R302.1.

Building Considerations

Stabilized, reinforced rammed-earth walls are an alternative to the prescriptive requirements of the IRC. In this case, the term "stabilized" refers to the fact that Portland cement is added to the soil to help bind it, increase durability, and improve its structural properties. "Reinforced" means that steel reinforcing bar will be used much like it would be with concrete. As such, this type of wall system will require extensive research to determine the structural properties of the soil materials used in the wall, proper ratio of cement to soil, fire-resistive properties, and more. Here, it is critical to discuss matters with the code official. Identifying specific information needed for approval upfront will improve the chances of incorporating the AMM into the project.

After extensive research, it is determined that there is no nationally recognized standard for rammed-earth construction; thus, it will be necessary to provide data based on independent research and published test results. A quick search on the Internet yields published reports from the Massachusetts Institute of Technology (MIT) and the University of Southern California School of Architecture, among others, that may prove useful.

It is imperative that these findings and others be shared with the code official at the earliest stages of the design so that it can be determined that they are satisfactory. Remember, she is under no obligation to accept the data or the AMM, so it will pay to do the homework. Assuming the code official has agreed to allow the AMM and that the specific design criteria are established, the design professional can set about the task of providing a complete, comprehensive design based on accepted engineering practice.

The interior walls will be framed with light-gauge steel framing, which is consistent with our desire to minimize the use of wood wherever possible. Section R603 of the IRC addresses prescriptive wall framing for steel frame walls, so it is relatively easy to incorporate steel into the sample project. The necessary steel thicknesses, connection details, and specifications are outlined in this section of the code.

CHAPTER SUMMARY

Chapter 6 of the IRC addresses the various code requirements for wall construction. It provides standards for materials and methods of installation for both wood and steel construction materials, masonry construction, and SIPs used in wall framing. Many of the materials covered in the chapter offer green alternatives or are actually regarded as green themselves. Following are highlights of the chapter.

- Chapter 6 of the IRC provides prescriptive code requirements for both wood and steel frame walls, masonry, SIPs, and ICF walls.
- Criteria for wall framing systems are provided in both text and tabulated form.
- Wood, steel, and concrete wall framing systems collect and distribute loads to supporting elements.
- The allowable height of a wood or steel wall stud is a function of several criteria, including loading conditions, size and spacing of the member, and grade of the wood or steel used.
- Stud height tables are provided for both wood and steel construction.
- Many of the applications available for wall framing are already "green" alternatives and can simply be chosen as an allowed option under the code.
- Some options such as I-joists are green alternatives to conventional construction when accepted as AMM.
- The manufacturer's installation guidelines for engineered wood products must be observed.

INTERIOR AND EXTERIOR WALL COVERINGS

This chapter corresponds to IRC Chapter 7, Wall Coverings

l e a r n i n g   o b j e c t i v e

To know and understand how interior and exterior wall coverings are generally regulated by the code. Also, to explore green options for these materials and to understand how the IRC requirements impact those choices once made.

IRC CHAPTER 7—OVERVIEW

Chapter 7 of the IRC provides requirements for interior and exterior wall coverings. The chapter begins with Section R701.1, Application, which requires the design and construction of all wall coverings to comply with the provisions of the chapter. This charging statement is important because it establishes that the provisions of the chapter apply to *all* buildings. So, this chapter is just as applicable to accessory buildings and garages as it is to occupied buildings such as dwellings and duplexes or any other building that falls under the scope of the IRC.

There is an important distinction to be made here with respect to wall coverings. The IRC makes no attempt to regulate decorative products and installations such as paint, wallpaper, tile work, cabinetry, millwork, or other similar finish work, nor does it regulate carpeting or other flooring materials. Section R105.2, Work Exempt from Permit, is clear that these items are exempt from the permit process. It is helpful to think of it this way: If the material in question is cosmetic or otherwise decorative in nature, then that item is beyond the scope of the IRC and is thus not subject to regulation.

Instead, the code is concerned with those wall coverings that are attached directly to the wall structure or wall system. So, for example, traditional interior wall coverings such as gypsum wallboard (drywall), plaster and wood veneer, or hardboard paneling are regulated in some way under the provisions of the IRC. Likewise, traditional exterior wall coverings such as wood siding, brick or stone veneer, and exterior plaster (stucco) are regulated. Wall coverings are important because they create the occupant's interior environment and provide important protection from the elements with a weather-resistant barrier outside the dwelling.

The two principal themes apparent in this IRC chapter are to identify appropriate materials standards and to prescribe methods of attachment. As with all construction materials, the IRC establishes specific standards to which both interior and exterior wall coverings must adhere. These standards attempt to ensure that products used in dwellings are safe and will function in the manner in which they were intended.

For many of the conventional interior and exterior finish materials, backing materials, and weather-resistive barriers addressed in the code, ASTM standards are referenced. In addition, standards from the American National Standards Institute (ANSI) in conjunction with the Hardwood Plywood and Veneer Association (HPVA) and the American Hardboard Association (AHA) are referenced for wood veneer and hardboard paneling, as are Cedar Shake and Shingle Bureau (CSSB) standards for wood shake and shingle installations.

In large part, however, the IRC provides criteria for the method of attachment for the various products used to cover walls. Specific sizes, types, and spacing of fasteners are prescribed for each of the interior and exterior wall coverings addressed in the code. The attachment criteria are provided in either text form or tables, such as Table R702.3.5 (see Figure 9-1). This table prescribes the method of attachment for gypsum wallboard products used in wall and ceiling applications, and provides criteria for those applications that use adhesives and those that do not.

GREEN OPTIONS

Like so many other elements addressed in the code, the green options available to someone considering what sorts of materials to use as wall coverings are largely a matter of one's perspective on just what "green" means in the first place. This is just as true for Chapter 7 of the IRC as it is for the other chapters we have reviewed. What direction an individual takes in the decision to use this green method or that is largely a matter of personal preference.

To some, "green" means the use of natural materials for wall coverings, such as wood products instead of man-made materials. In these cases, wood siding or shingles might be the appropriate choice and the IRC provides criteria specifically for those materials. To others, "green" means exactly the opposite; the preference here is to use materials that are manufactured by man in an effort to reduce demands on natural resources. In these cases, fiber cement or vinyl siding might be the appropriate choices and, again, the IRC provides criteria for these materials.

Some who are planning a green dwelling prefer to use materials that create a pleasing interior, one that "feeds the soul" and creates a harmonious living space. They prefer materials that have a natural, inviting ambience. Indoor air quality is an issue for others, thus some prefer products that are free of formaldehydes or that do not use adhesives. For the majority, going green is some combination of all of these factors to one degree or another.

Given that there are so many variables affecting the decisions to be made in the choosing of materials for wall coverings, this text makes no attempt to prescribe which of these options is the most green. The reader decides which option is most meaningful and then chooses a course of action based on the requirements of the IRC and the suggestions provided here.

Figure 9-1

Table R702.3.5 Minimum thickness and application of gypsum board.

TABLE R702.3.5
MINIMUM THICKNESS AND APPLICATION OF GYPSUM BOARD

THICKNESS OF GYPSUM BOARD (inches)	APPLICATION	ORIENTATION OF GYPSUM BOARD TO FRAMING	MAXIMUM SPACING OF FRAMING MEMBERS (inches o.c.)	MAXIMUM SPACING OF FASTENERS (inches) Nails[a]	MAXIMUM SPACING OF FASTENERS (inches) Screws[b]	SIZE OF NAILS FOR APPLICATION TO WOOD FRAMING[c]
Application without adhesive						
$^3/_8$	Ceiling[d]	Perpendicular	16	7	12	13 gage, $1^1/_4$" long, $^{19}/_{64}$" head; 0.098" diameter, $1^1/_4$" long, annular-ringed; or 4d cooler nail, 0.080" diameter, $1^3/_8$" long, $^7/_{32}$" head.
	Wall	Either direction	16	8	16	
$^1/_2$	Ceiling	Either direction	16	7	12	13 gage, $1^3/_8$" long, $^{19}/_{64}$" head; 0.098" diameter, $1^1/_4$" long, annular-ringed; 5d cooler nail, 0.086" diameter, $1^5/_8$" long, $^{15}/_{64}$" head; or gypsum board nail, 0.086" diameter, $1^5/_8$" long, $^9/_{32}$" head.
	Ceiling[d]	Perpendicular	24	7	12	
	Wall	Either direction	24	8	12	
	Wall	Either direction	16	8	16	
$^5/_8$	Ceiling	Either direction	16	7	12	13 gage, $1^5/_8$" long, $^{19}/_{64}$" head; 0.098" diameter, $1^3/_8$" long, annular-ringed; 6d cooler nail, 0.092" diameter, $1^7/_8$" long, $^1/_4$" head; or gypsum board nail, 0.0915" diameter, $1^7/_8$" long, $^{19}/_{64}$" head.
	Ceiling[e]	Perpendicular	24	7	12	
	Wall	Either direction	24	8	12	
	Wall	Either direction	16	8	16	
Application with adhesive						
$^3/_8$	Ceiling[d]	Perpendicular	16	16	16	Same as above for $^3/_8$" gypsum board
	Wall	Either direction	16	16	24	
$^1/_2$ or $^5/_8$	Ceiling	Either direction	16	16	16	Same as above for $^1/_2$" and $^5/_8$" gypsum board, respectively
	Ceiling[d]	Perpendicular	24	12	16	
	Wall	Either direction	24	16	24	
Two $^3/_8$ layers	Ceiling	Perpendicular	16	16	16	Base ply nailed as above for $^1/_2$" gypsum board; face ply installed with adhesive
	Wall	Either direction	24	24	24	

For SI: 1 inch = 25.4 mm.

a. For application without adhesive, a pair of nails spaced not less than 2 inches apart or more than 2½ inches apart may be used with the pair of nails spaced 12 inches on center.

b. Screws shall be in accordance with Section R702.3.6. Screws for attaching gypsum board to structural insulated panels shall penetrate the wood structural panel facing not less than $^7/_{16}$ inch.

c. Where cold-formed steel framing is used with a clinching design to receive nails by two edges of metal, the nails shall be not less than $^5/_8$ inch longer than the gypsum board thickness and shall have ringed shanks. Where the cold-formed steel framing has a nailing groove formed to receive the nails, the nails shall have barbed shanks or be 5d, 13½ gage, $^{15}/_{64}$ inch head for ½-inch gypsum board; and 6d, 13 gage, $1^7/_8$ inches long, $^{15}/_{64}$-inch head for $^5/_8$-inch gypsum board.

d. Three-eighths-inch-thick single-ply gypsum board shall not be used on a ceiling where a water-based textured finish is to be applied, or where it will be required to support insulation above a ceiling. On ceiling applications to receive a water-based texture material, either hand or spray applied, the gypsum board shall be applied perpendicular to framing. When applying a water-based texture material, the minimum gypsum board thickness shall be increased from $^3/_8$ inch to ½ inch for 16-inch on center framing, and from ½ inch to $^5/_8$ inch for 24-inch on center framing or ½-inch sag-resistant gypsum ceiling board shall be used.

e. Type X gypsum board for garage ceilings beneath habitable rooms shall be installed perpendicular to the ceiling framing and shall be fastened at maximum 6 inches o.c. by minimum $1^7/_8$ inches 6d coated nails or equivalent drywall screws.

INTERIOR WALL COVERINGS

As the name implies, interior wall coverings are those materials that cover the walls inside the dwelling. They primarily form the interior living space or, essentially, what the occupants of a dwelling see every day. Thus, they are important for a variety of reasons. From a code standpoint, interior wall coverings provide a barrier between the interior space and the wall system including the insulation, wiring, and plumbing. More important, however, interior wall coverings create and define the interior living environment. Because we spend so much of our lives indoors, it is imperative that the indoor environment be safe and comfortable, and afford suitable protection.

Interior wall coverings are required to meet the flame-spread and smoke-developed indices identified in Section R302.9, which you will recall are 200 and 450, respectively. If you are considering the use of an AMM, it will need to meet these general requirements, and be at least the equivalent of those conventional materials specified in the code.

Section R702.2 Interior Plaster

Although not as common as gypsum wallboard in modern residential construction, plaster is still used in some areas. It is used extensively in older homes during renovation and restoration projects and for general wall repair. Plaster can be applied over a variety of supporting materials including gypsum, metal, or wire lath. It can also be applied directly to the face of concrete and masonry walls as a finish surface, and in some cases installed as the finish coat over gypsum sheathing or gypsum veneer.

The IRC establishes standards for both gypsum plaster and Portland cement plaster. Table R702.1(1) identifies both minimum and maximum thicknesses of plaster depending on a number of variables, including the type of base over which the plaster is applied and the type of plaster used (see Figure 9-2). Additionally, the IRC establishes the appropriate proportions of raw materials to be used in the plaster mix, again determined by both the type of plaster to be used and other variables such as type of lath used, whether the plaster will be installed in a two-coat or three-coat system, and whether the plaster is to be used in the first, second, or third coats.

Section R702.2 completes the code requirements for plaster applications, providing the support spacing requirements for gypsum lath, based on the lath thickness as well as the criteria identified above.

Intent

The intent of this IRC section is to establish minimum materials and installation standards for interior plaster wall coverings used in dwellings. The code further intends that materials used in the application of interior plaster meet appropriate ASTM standards for quality and installation. This code section also intends that interior plaster wall coverings be attached securely to the supporting structure.

Figure 9-2

Table R702.1(1) Thickness of plaster.

TABLE R702.1(1)
THICKNESS OF PLASTER

PLASTER BASE	FINISHED THICKNESS OF PLASTER FROM FACE OF LATH, MASONRY, CONCRETE (inches)	
	Gypsum Plaster	Cement Plaster
Expanded metal lath	$^5/_8$, minimum[a]	$^5/_8$, minimum[a]
Wire lath	$^5/_8$, minimum[a]	$^3/_4$, minimum (interior)[b] $^7/_8$, minimum (exterior)[b]
Gypsum lath[g]	$^1/_2$, minimum	$^3/_4$, minimum (interior)[b]
Masonry walls[c]	$^1/_2$, minimum	$^1/_2$, minimum
Monolithic concrete walls[c, d]	$^5/_8$, maximum	$^7/_8$, maximum
Monolithic concrete ceilings[c, d]	$^3/_8$, maximum[e]	$^1/_2$, maximum
Gypsum veneer base[f, g]	$^1/_{16}$, minimum	$^3/_4$, minimum (interior)[b]
Gypsum sheathing[g]	—	$^3/_4$, minimum (interior)[b] $^7/_8$, minimum (exterior)[b]

COPYRIGHT 2009 IRC®

For SI: 1 inch = 25.4 mm.

a. When measured from back plane of expanded metal lath, exclusive of ribs, or self-furring lath, plaster thickness shall be $^3/_4$ inch minimum.
b. When measured from face of support or backing.
c. Because masonry and concrete surfaces may vary in plane, thickness of plaster need not be uniform.
d. When applied over a liquid bonding agent, finish coat may be applied directly to concrete surface.
e. Approved acoustical plaster may be applied directly to concrete or over base coat plaster, beyond the maximum plaster thickness shown.
f. Attachment shall be in accordance with Table R702.3.5.
g. Where gypsum board is used as a base for cement plaster, a water-resistive barrier complying with Section R703.2 shall be provided.

Green Decisions and Limitations

Shall I use a green alternative to conventional gypsum or Portland cement plaster? How will the requirements of the IRC impact my decision?

Options

Where the requirements of IRC Section R702.2 apply, consider the following green alternatives.

Tadelakt plaster: This lime plaster product comes from the Marrakech region of Morocco. It is purported to be one of the oldest forms of lime plaster and has been used for centuries in and on buildings throughout this region. It is naturally durable and contains a relatively high clay content compared to other lime plasters. Tadelakt plaster does not contain Portland cement. Like other earth plasters, this product does not encourage mold growth. This natural product can be used both on the interior and the exterior of a dwelling but is most commonly used indoors (see Figure 9-3).

It is considered by some to be difficult to apply in the traditional method, but the results are considered worth the extra effort. Like most plasters, Tadelakt can

Figure 9-3

Tadelakt plaster in a bathtub area. This wall finish is a traditional Moroccan finish, but is now used more frequently in other parts of the world.

SOURCE: TIERRAFINO AMSTERDAM

Why It's Green

Alternative plastering materials such as earth and clay plaster are green because they do not contain volatile organic compounds (VOCs) and do not produce off-gasses as they cure, which can reduce indoor air pollution and improve air quality. Also, these materials do not contain added chemicals such as formaldehyde. Alternative plastering materials are also minimally processed, meaning they contain less embodied energy. In some areas, earth and clay plasters are made from local materials, which can reduce the environmental impact of shipping materials over long distances. Earth-based plasters do not encourage mold growth.

be colored with pigments to create a variety of looks and textures. There are some synthetic Tadelakt products available as well, which might be easier to obtain in your region.

Italian plaster: Lime plasters from various regions in Italy have been used for centuries as well. These plasters have been used in some of the world's most important and visually stunning buildings. Limestone produced in the Venezia region of Italy is among the purest in the world and highly prized. Many of the plasters use marble dust, crystal, mica, and other additives to create a particular look and finish (see Figure 9-4).

Italian plasters can be finished in a variety of ways. They can be colored with pigments and display various finishes: Stucco Romano, which includes marble dust and resin; Veneziano, which creates subtle, contrasting layers of color; and

Figure 9-4

An example of
Italian plaster in the
prontofresco style;
muschio padano finish.

Fresconova, a pure lime plaster that is used in much of the restoration work done in Italy because it so closely matches the traditional plastering methods used throughout the centuries.

Tadelakt and Italian lime plasters are but a few of the many types of plasters available. With some research, the reader may find additional earth plasters available regionally, with the added benefit of reduced environmental impacts due to transporting other materials over long distances.

Section R702.3 Gypsum Wallboard

Gypsum wallboard is, by far, the most commonly used material for interior wall coverings in the United States and Canada today. According to the U.S. Gypsum Association, market demands in these two countries alone necessitate the manufacture of some 30 billion square feet annually. Gypsum wallboard panels are relatively lightweight, easy to work, and cover large areas in a short amount of time. These characteristics make it an extremely popular choice among today's builders, as it is used extensively in residential and commercial construction (see Figure 9-5).

Gypsum wallboard, commonly called drywall and plasterboard, is made up of a noncombustible core of gypsum which is covered on all sides with a thin paper coating. Gypsum wallboard panels come in a variety of thicknesses; however, 0.5-inch (13-mm) and 0.63-inch (16-mm) thick panels are most commonly encountered in the average dwelling. Some gypsum wallboard products have enhanced

Figure 9-5

Gypsum wallboard stacked and ready for application to the walls of a dwelling.

COURTESY DELMAR/CENGAGE LEARNING

fire-resistive properties and are referred to as Type X panels when tested to meet the ASTM C 36 standard. Type X panels have fiberglass or other noncombustible fiber reinforcement added to the gypsum slurry when the panels are formed, to strengthen the panels and enhance their fire-resistive properties.

Gypsum wallboard is used on both walls and ceilings in dwellings. Table R702.3.5 provides the criteria by which the wallboard is to be installed (see Figure 9-1). Requirements for fastener type, size, and spacing vary by the manner in which the panels are applied (walls or ceilings), the spacing of the supporting framing members, and the thickness of the panels themselves. Additionally, Table R702.3. 5 provides criteria for applications that use adhesives and those that do not.

Like gypsum plaster, wallboard is required to comply with specific ASTM standards, which govern the panels themselves, the fasteners, and methods of installation. Gypsum wallboard panels can be supported by wood or metal framing and also can be used to cover the foam plastic used in insulating concrete form (ICF) walls as required by Sections R316.4 and R611.4.2 of the IRC.

Interior gypsum wallboard is prohibited in areas where the material will be exposed to the weather, to water, or to high humidity such as in saunas and swimming pool enclosures. While the gypsum that forms the core of the panel is inert and relatively stable, the paper facing covering the core will absorb moisture and can degrade or foster mold growth when installed in high-moisture areas. In these applications, water-resistant backing board that conforms to ASTM C 1396, C 1178, or C 1278 may be allowed and can also serve as the base for the application of ceramic tile or other nonabsorbent finishes. In these cases, the product must be installed per Section R702.3.8 of the IRC and per the manufacturer's installation instructions. Fiber cement, fiber-reinforced gypsum backers, and other similar materials conforming to ASTM C 1288, C 1325, C 1178, or C 1278 may also be used as a backing for ceramic tile.

Intent

The intent of this IRC section is to establish minimum standards for the materials and methods of installation for gypsum wallboard panels used as interior wall coverings. The code further intends that gypsum wallboard products conform to ASTM standards for product quality and method of installation. The code also intends that gypsum wallboard products are securely fastened to the supporting structure.

Why It's Green

Mold-resistant gypsum wall-board incorporates a biocide during the manufacturing process to discourage mold growth. An interior environment that does not foster mold growth is generally regarded as having better indoor air quality, among other health benefits. Where concerns arise over the use of biocides, say due to extreme chemical sensitivity, glass mat-faced gypsum board and the use of other special facings offer additional mold-resistant options.

Green Decisions and Limitations

Shall I use mold-resistant gypsum wallboard in my green dwelling? How will the requirements of the IRC impact this decision?

Gypsum Wallboard Option

In response to concerns in recent years over the health concerns related to mold growth inside dwellings, some manufacturers of gypsum wall board panels have developed products that are touted to be mold resistant (see Figure 9-6). In some cases, the manufacturer treats the paper facing on both sides of the gypsum wallboard panel with a biocide to reduce the chances of mold growth. In other cases, the product manufacturer adds a biocide to the gypsum slurry as it is made, or uses products other than paper, such as glass mats, to reduce the likelihood of mold growth.

There is an important point to be made here: The term "mold resistant" is not synonymous with "mold proof." Under the right conditions, it is possible to grow mold on virtually any surface. To be regarded as mold resistant, gypsum wall board products are required to be manufactured to ASTM D 3273, Standard Test for Resistance to Growth of Mold on the Surface of Interior Coatings in an Environmental Chamber. Mold-resistant gypsum wallboard products can be installed in generally the same way as conventional products, making their use in residential construction fairly easy. Many of these products have evaluation reports available.

Figure 9-6

Mold-resistant gypsum board applied to the inside face of the exterior walls in a remodeling project.

Be sure to check these and the manufacturer's recommendations prior to installation and make sure the code official knows of your intent to use the product.

EXTERIOR WALL COVERINGS

Exterior wall coverings form the primary weather barrier for a dwelling. It is essential that products used in these applications perform as intended and that they are reasonably durable. Section R703.1 requires exterior wall coverings to provide the building with the weather-resistant exterior wall envelope. The code requires flashings and other means to be used to prevent water from accumulating inside the **wall assembly** (all wall components including studs, sheathing, insulation, interior and exterior wall coverings, papers, etc.) through the use of a water-resistant barrier behind the exterior veneer.

Furthermore, the IRC requires a means of drainage for water that enters the wall assembly. Why is this required? Ask most experts about water infiltration and they will tell you that it is not a question of "if" water will enter the assembly, but "when." Water can enter the wall assembly from the outside in the form of a liquid or as water vapor; and it can be generated from inside the wall due to plumbing leaks. Water vapor entering the wall from the exterior or from the interior of the dwelling (from cooking, bathing, etc.) can condense inside the wall assembly, all of which can lead to serious moisture problems.

Water problems occur for a variety of reasons. Over time, construction materials dry out or shrink and buildings can settle. Both conditions can allow water to migrate inside the wall assembly. Caulks and other sealants break down over time due to exposure to the elements and require regular maintenance. Where this maintenance does not occur in a timely manner, water can migrate through these openings as well. Flashing around windows or doors is sometimes forgotten during construction, meaning that water will inevitably enter the wall assembly. When it does, the water must have a means of drainage so that decay and mold do not occur.

The IRC prescribes methods of attachment for all common forms of exterior wall coverings. Table R703.4 provides attachment criteria for type, size, and spacing of fasteners like other tables in this chapter, but it also provides important criteria for the treatment of joints, minimum thicknesses of materials to be used, and important footnotes that affect the installation requirements for many of the materials covered (see Figure 9-7).

TAKE NOTE!

Footnotes appear after virtually every table used in the code. Footnotes provide critical qualifying information that sometimes waives tabulated requirements or further restricts the information provided in the table. Be sure to know and understand how footnotes modify the requirements of the code. If you are unsure, ask the code official in your area for guidance.

Figure 9-7

Table R703.4 Weather-resistant siding attachment and minimum thickness.

TABLE R703.4
WEATHER-RESISTANT SIDING ATTACHMENT AND MINIMUM THICKNESS

SIDING MATERIAL		NOMINAL THICKNESS[a] (inches)	JOINT TREATMENT	TYPE OF SUPPORTS FOR THE SIDING MATERIAL AND FASTENERS[b, c, d]						
				WATER-RESISTIVE BARRIER REQUIRED	Wood or wood structural panel sheathing	Fiberboard sheathing into stud	Gypsum sheathing into stud	Foam plastic sheathing into stud	Direct to studs	Number or spacing of fasteners
Horizontal aluminum[e]	Without insulation	0.019[f]	Lap	Yes	0.120 nail 1½" long	0.120 nail 2" long	0.120 nail 2" long	0.120 nail[y]	Not allowed	Same as stud spacing
		0.024	Lap	Yes	0.120 nail 1½" long	0.120 nail 2" long	0.120 nail 2" long	0.120 nail[y]	Not allowed	
	With insulation	0.019[f]	Lap	Yes	0.120 nail 1½" long	0.120 nail 2½" long	0.120 nail 2½" long	0.120 nail[y]	0.120 nail 1½" long	
Anchored veneer: brick, concrete, masonry or stone		2	Section R703	Yes	See Section R703 and Figure R703.7[g]					
Adhered veneer: concrete, stone or masonry[w]		—	Section R703	Yes Note w	See Section R703.6.1[g] or in accordance with the manufacturer's instructions.					
Hardboard[k] Panel siding-vertical		7/16	—	Yes	Note m	Note m	Note m	Note m	Note m	6" panel edges 12" inter. sup.[n]
Hardboard[k] Lap-siding-horizontal		7/16	Note p	Yes	Note o	Note o	Note o	Note o	Note o	Same as stud spacing 2 per bearing
Steel[h]		29 ga.	Lap	Yes	0.113 nail 1¾" Staple–1¾"	0.113 nail 2¾" Staple–2½"	0.113 nail 2½" Staple–2¼"	0.113 nail[y] Staple[y]	Not allowed	Same as stud spacing
Particleboard panels		3/8–½	—	Yes	6d box nail (2" × 0.099")	6d box nail (2" × 0.099")	6d box nail (2" × 0.099")	box nail[v]	6d box nail (2" × 0.099"), 3/8 not allowed	6" panel edge, 12" inter, sup.h
		5/8	—	Yes	6d box nail (2" × 0.099")	8d box nail (2½" × 0.113")	8d box nail (2½" × 0.113")	box nail[v]	6d box nail (2" × 0.099")	
Wood structural panel siding[i] (exterior grade)		3/8–1/2	Note p	Yes	0.099 nail–2"	0.113 nail–2½"	0.113 nail–2½"	0.113 nail[v]	0.099 nail–2"	6" panel edges, 12" inter, sup.
Wood structural panel lapsiding		3/8–½	Note p Note x	Yes	0.099 nail–2"	0.113 nail–2 ½"	0.113 nail–2½"	0.113 nail[x]	0.099 nail–2"	8" along bottom edge
Vinyl siding[l]		0.035	Lap	Yes	0.120 nail (shank) with a 0.313 head or 16 gauge staple with 3/8 to ½-inch crown[y,z]	0.120 nail (shank) with a 0.313 head or 16 gage staple with 3/8 to ½-inch crown[y]	0.120 nail (shank) with a 0.313 head or 16 gage staple with 3/8 to ½-inch crown[y]	0.120 nail (shank) with a 0.313 head per Section R703.11.2	Not allowed	16 inches on center or specified by the manufacturer instructions or test report
Wood[j] rustic, drop		3/8 Min	Lap	Yes	Fastener penetration into stud-1"				0.113 nail–2½" Staple–2"	Face nailing up to 6" widths, 1 nail per bearing; 8" widths and over, 2 nails per bearing
Shiplap		19/32 Average	Lap	Yes						
Bevel		7/16								
Butt tip		3/16	Lap	Yes						
Fiber cement panel siding[q]		5/16	Note q	Yes Note u	6d common corrosion-resistant nail[r]	6d common corrosion-resistant nail[r]	6d common corrosion-resistant nail[r]	6d common corrosion resistant (12" × 0.113") nail[r,v]	4d common corrosion resistant nail[r]	6" o.c. on edges, 12" o.c. on intermed. studs
Fiber cement lap siding[s]		5/16	Note s	Yes Note u	6d common corrosion-resistant nail[r]	6d common corrosion-resistant nail[r]	6d common corrosion-resistant nail[r]	6d common corrosion-resistant (12" × 0.113") nail[r,v]	6d common corrosion-resistant nail or 11 gage roofing nail[r]	Note t

TABLE R703.4—continued
WEATHER-RESISTANT SIDING ATTACHMENT AND MINIMUM THICKNESS

For SI: 1 inch = 25.4 mm.

a. Based on stud spacing of 16 inches on center where studs are spaced 24 inches, siding shall be applied to sheathing approved for that spacing.

b. Nail is a general description and shall be T-head, modified round head, or round head with smooth or deformed shanks.

c. Staples shall have a minimum crown width of $^7/_{16}$-inch outside diameter and be manufactured of minimum 16 gage wire.

d. Nails or staples shall be aluminum, galvanized, or rust-preventative coated and shall be driven into the studs for fiberboard or gypsum backing.

e. Aluminum nails shall be used to attach aluminum siding.

f. Aluminum (0.019 inch) shall be unbacked only when the maximum panel width is 10 inches and the maximum flat area is 8 inches. The tolerance for aluminum siding shall be +0.002 inch of the nominal dimension.

g. All attachments shall be coated with a corrosion-resistant coating.

h. Shall be of approved type.

i. Three-eighths-inch plywood shall not be applied directly to studs spaced more than 16 inches on center when long dimension is parallel to studs. Plywood $^1/_2$-inch or thinner shall not be applied directly to studs spaced more than 24 inches on center. The stud spacing shall not exceed the panel span rating provided by the manufacturer unless the panels are installed with the face grain perpendicular to the studs or over sheathing approved for that stud spacing.

j. Wood board sidings applied vertically shall be nailed to horizontal nailing strips or blocking set 24 inches on center. Nails shall penetrate 1½ inches into studs, studs and wood sheathing combined or blocking.

k. Hardboard siding shall comply with CPA/ANSI A135.6.

l. Vinyl siding shall comply with ASTM D 3679.

m. Minimum shank diameter of 0.092 inch, minimum head diameter of 0.225 inch, and nail length must accommodate sheathing and penetrate framing 1½ inches.

n. When used to resist shear forces, the spacing must be 4 inches at panel edges and 8 inches on interior supports.

o. Minimum shank diameter of 0.099 inch, minimum head diameter of 0.240 inch, and nail length must accommodate sheathing and penetrate framing 1½ inches.

p. Vertical end joints shall occur at studs and shall be covered with a joint cover or shall be caulked.

q. See Section R703.10.1.

r. Fasteners shall comply with the nominal dimensions in ASTM F 1667.

s. See Section R703.10.2.

t. Face nailing: one 6d common nail through the overlapping planks at each stud. Concealed nailing: one 11 gage 1½ inch long galv. roofing nail through the top edge of each plank at each stud.

u. See Section R703.2 exceptions.

v. Minimum nail length must accommodate sheathing and penetrate framing 1½ inches.

w. Adhered masonry veneer shall comply with the requirements of Section R703.6.3 and shall comply with the requirements in Sections 6.1 and 6.3 of ACI 530/ASCE 5/TMS-402.

x. Vertical joints, if staggered shall be permitted to be away from studs if applied over wood structural panel sheathing.

y. Minimum fastener length must accommodate sheathing and penetrate framing .75 inches or in accordance with the manufacturer's installation instructions.

z. Where approved by the manufacturer's instructions or test report siding shall be permitted to be installed with fasteners penetrating not less than .75 inches through wood or wood structural sheathing with or without penetration into the framing.

Section R703.3 Wood, Hardboard, and Wood Structural Panel Siding; Section R703.5 Wood Shakes and Shingles

These IRC sections provide the code requirements for the installation of various types of wood products used as exterior wall coverings. The first code section covers horizontal wood siding (e.g., cedar siding), hardboard (composite wood) products such as horizontal or panel siding made from wood fibers and binders (e.g., particleboard), and wood structural panel siding (wood panels that are, at once, the exterior wall covering and the structural sheathing necessary to brace the structure from lateral forces). The latter code section provides requirements for the use of wood shakes and shingles, another popular form of exterior wall covering.

Wood products in general have been used for over two centuries in the United States. Properly maintained exterior wall coverings made of wood can last for many decades. The key words here are "properly maintained," as all wood products are subject to degradation over time from the effects of rain, snow, wind, and sunlight. Improper maintenance can rapidly accelerate the degradation process. Wood siding, shakes, and shingles can be painted or stained to create a charming, weather-tight exterior, which remains a popular choice today (see Figure 9-8).

The code prescribes methods of attachment and joint treatments in general for these types of wood products in Table R703.4 (refer to Figure 9-7). It also prescribes *specific* joint treatment requirements for panel siding and horizontal siding in Sections R703.3.1 and R703.3.2 and includes lap requirements for these products as well.

Wood shakes and shingles are addressed in Section R703.5.1, which prescribes the method of attachment, limitations on exposure to weather, and other pertinent information necessary for a code-compliant installation. Code criteria for method

Figure 9-8

Wood shingles applied to the exterior of a dwelling, in this case a dormer. Wood shingles can be stained, sealed, or left unfinished making them a flexible, durable choice.

of attachment varies based on geographic locations, as prescribed in Section R703.4, which requires the attachment of exterior wall coverings to be designed where basic wind speeds exceed 110 mph (49.2 m/s) (such as coastal regions).

Intent

The intent of this IRC section is to ensure that wood and wood-based products used as exterior wall coverings are suitable for use in an exterior application. The code further intends that exterior wall coverings be properly attached to the supporting structure.

Green Decisions and Limitations

Are there green alternatives to wood products for use as exterior wall coverings?

How does the IRC impact this decision?

Options

When selecting exterior wall finishes that will meet the requirements of IRC Section R703, consider the following green options.

Stone and masonry veneer: Veneers such as stone and brick create a classic and durable exterior. They are naturally weather resistant and provide insulative properties to the exterior wall assembly (see Figure 9-9). Because they are wood-free alternatives, stone and brick reduce demands for lumber products, which contribute to a more sustainable method of construction. They are suitable in all climates and can be used indoors as well.

Stone and masonry veneers are specifically allowed under the provisions of the IRC. Requirements for the installation of these construction materials are provided in Table R703.4 (refer to Figure 9-7) and in Section R703.7. Because stone and masonry veneers are heavy materials, they must be considered in the overall design of the structure at the earliest stages of the design. Footings and foundations must be wide enough to accommodate the width and weight of the stone or masonry, and where these materials will be supported by wood framing, the additional weights must be taken into account when sizing the supporting members.

Why It's Green

Stone or masonry veneers are a green alternative to the use of wood products as exterior wall coverings because they reduce demand for wood products, which can potentially preserve forests. Stone and masonry veneers are naturally durable materials and require considerably less maintenance than wood products. Like stone or masonry, fiber cement siding materials also reduce demands for wood products.

TAKE NOTE!

Where stone and masonry veneers are used indoors, they are permitted to be supported by wood or light-gauge steel frame floors only if the additional loads from the stone or masonry have been considered in the design of the floor. Prescriptive floor framing span tables provided in Chapter 5 of the IRC cannot be used in these applications due to the additional loading from these veneers.

Figure 9-9

Brick veneer applied to a dwelling. Brick is naturally durable and weather resistant, making it a popular choice for green dwellings.

COURTESY DELMAR/CENGAGE LEARNING

Like many other code requirements, provisions for stone and masonry also vary by geographic region. In areas that are prone to high seismic activity, height limitations for the veneers apply, special requirements are established for reinforcing steel, and attachment methods are defined. These and the preceding requirements are examples of where specific code requirements impact the green decision both before it is made and after. It is best to decide to use these materials as early as possible in the planning and design stages.

The IRC provides several construction details that are useful in the design and construction of stone and masonry veneer applied over wood or steel framing.

Fiber cement siding: Some siding materials, such as fiber cement siding, are man-made materials engineered to be durable and meet all code requirements for exterior wall coverings. Fiber cement siding products are made from sand, cement, and cellulose (wood pulp) fibers. The products have a woodlike appearance and can be manufactured in a variety of different configurations, such as horizontal siding, panel siding, vertical siding, shingles, and soffit and trim material (see Figure 9-10). Fiber cement products are workable with standard construction tools. Although fiber cement utilizes wood to some degree, overall

Figure 9-10

Fiber cement siding and trim applied to the exterior of a dwelling.

COURTESY OF JAMES HARDIE

TAKE NOTE!

A word to the wise regarding manufactured products used as wall and ceiling coverings: Most manufacturers have specific installation guidelines and instructions for their products that must be closely followed in order to ensure that the materials and systems will perform as intended. The IRC requires a manufacturer's installation criteria to be followed, in addition to any code-imposed requirements. More important, products that are improperly installed can limit or void a manufacturer's warranties! Some manufacturers even require their products to be installed by factory-trained installers. Be sure you know and understand what is required by the manufacturer to protect your investment!

these products use less than products that are made from 100% wood materials such as wood shakes and shingles.

Like stone and masonry veneer, fiber cement siding is allowed outright under the code. Its methods of attachment are as prescribed in Table R703.4 (refer to Figure 9-7). In addition to the code-specified method of connection, manufacturers' criteria apply as well.

CHAPTER 9 APPLIED TO THE SAMPLE PROJECT

Chapter 9 pertains to interior and exterior wall coverings. It addresses the use of plaster and gypsum board as interior wall coverings, as well as the use of brick and stone veneer, wood shakes, and other exterior wall coverings. The following analysis will demonstrate the application of some IRC code provisions for floors with consideration given for the proposed scope of work and our design philosophy.

Site Considerations

The location of the exterior wall in relation to the exterior property line requires the wall to be rated for 1-hour fire resistance. The other exterior walls are unaffected due to site considerations. Wall coverings (i.e., weather barriers) are not typically rated for fire resistance. Instead, the wall is protected underneath the weather barrier with various materials that have been tested for fire resistance. This is moot in our case, due to the use of rammed earth for the exterior walls which will be left in their natural state in our sample project.

Building Considerations

As stated, the rammed-earth exterior walls will be left in their natural state on both the exterior and the interior of our sample dwelling. Interior wall finishes (in this case, the earth face of the wall) must meet the requirements for flame spread and smoke developed as outlined in the code. Recall that those indices are 200 and 450, respectively, as required by Section R302.9 of the IRC. Thus, we will also need to ensure in our AMM that we find suitable data in our research to address these important criteria. Given that earth is essentially noncombustible, we would expect to find that these indices are near zero; however, it is not adequate to assume that this is true. We must demonstrate to a reasonable degree that this is in fact the case.

Interior wall coverings will be provided through the use of gypsum board, installed per the requirements of IRC Section R702.3 and per the installation criteria in Table R702.3.5. Recall that we will utilize steel framing for the walls and roof framing, thus the provisions of IRC Section R702.3.3, Cold-Formed Steel Framing, will also apply.

CHAPTER SUMMARY

Chapter 7 of the IRC covers interior and exterior wall coverings. It provides standards for materials and methods of installation for a variety of commonly used construction materials. Many of the materials covered in the chapter offer green alternatives or are actually green themselves. Following are highlights of the chapter.

- The IRC regulates interior and exterior wall coverings that are attached to the framing members.
- Criteria are provided in the text of the code as well as in tabulated form. Methods of attachment and type, size, and spacing of fasteners are identified.
- The IRC does not regulate decorative or cosmetic finish materials such as paint, wallpaper, millwork, cabinetry, and the like.
- The manufacturer's installation criteria must be followed as directed by the IRC to ensure a code-compliant installation and to preserve the manufacturer's warranty.
- Green alternatives to conventional materials are available, the selection of which depends on how one defines "green" construction.
- Some traditional construction materials are regarded as "green" and are specifically allowed under the provisions of the IRC.

ROOF AND CEILING FRAMING

This chapter corresponds to IRC Chapter 8, Roof-Ceiling Construction

learning objective

To know and understand how roofs and ceilings are regulated under the provisions of the IRC and to explore options for green roof-ceiling construction.

IRC CHAPTER 8—OVERVIEW

Chapter 8 of the International Residential Code regulates the construction of roofs and ceilings for dwellings. Like walls, roofs and ceilings shield us from the elements by forming a protective "lid" over our heads. Roofs provide structural support to resist the weight of accumulated snow, ice, and hail as these elements build up during inclement weather. They also divert rain and runoff from snow and ice melt to the ground below.

Roofs are made up of sloping structural members called rafters. Ceilings are made up of both horizontal and sloping structural members called ceiling joists. The ends of rafters and ceiling joists are typically supported by walls but can also be supported by beams, ledgers, and other structural elements within the building. Roofs can also be formed out of wood roof trusses, which are engineered, manufactured framing systems formed out of smaller lumber components, such as 2 inch by 3 inch (51 mm by 76 mm), 2 inch by 4 inch (51 mm by 102 mm), and 2 inch by 6 inch (51 mm by 152 mm), and larger components that are fastened with special connectors (see Figure 10-1). Wood roof trusses come in many configurations and can span long distances, despite the relatively small size of the framing members used in their fabrication.

Roofs and ceilings are similar to floors in terms of structural behavior and have an equally important role in the structural system for that reason. Like floors, roofs support gravity loads and they also distribute lateral forces to the supporting walls which, in turn, distribute those forces into other structural elements and so on. The IRC establishes code criteria for roofs and ceilings to ensure that loads are safely and effectively transferred to the resisting elements.

Figure 10-1

Manufactured roof trusses used to frame the roof-ceiling system in a dwelling. Note the use of 2 inch by 4 inch (51 mm by 102 mm) lumber and the metal plates used at the joints to fasten the wood components.

COURTESY DELMAR/CENGAGE LEARNING

Unlike floors, however, roofs receive their loads from different sources. In terms of gravity loads, where floors receive their loads from the contents and people inside a dwelling, roofs must support the weight of accumulated snow and ice, roof-mounted equipment such as air conditioner compressors or heat pumps, and, intermittently, the weight of service personnel to maintain equipment and the roof itself.

As we have seen in other chapters, some of these loads fluctuate from region to region. Snow loads, for example, vary considerably in different parts of the United States and Canada and the IRC is, of course, equipped to address these regional variances just as it is with other geographically influenced design criteria. In some areas, snow is virtually nonexistent and in other areas, such as the Mt. Hood region in Oregon, design ground snow loads regularly reach 280 pounds per square foot (13.407 kPa).

Ceilings, similarly, support gravity loads that may include the weights of items stored in an attic (holiday decorations, for example) or the weight of attic-mounted HVAC equipment; and they must support the weight of people who periodically enter the attic to access stored goods or to service equipment (see Figure 10-2). Some attics, such as those with low head room clearances, may never support anything other than the dead load weights of the ceiling framing and insulation. Although they can be designed to collect and distribute lateral forces, ceilings are not typically used for this purpose. It is easy to see that the loading conditions for roofs and ceilings can vary widely depending on the specific conditions to which they are exposed.

By now, it should be apparent that, like other structural components, rafter and ceiling joists support two primary types of load—gravity loads and lateral forces— just as their floor framing counterparts do. There is, however, another significant

Figure 10-2

An attic with a small storage floor installed on top of the ceiling joists.

COURTESY OF ATTICRAFT

loading condition that affects rafters and ceiling joists uniquely that must be taken into account in roof construction.

Prescriptively framed rafters are installed at an upward slope and the ever-present force of gravity acting upon them tends to pull them in a downward direction. The upper ends of each rafter are aligned and balanced with one another against the ridge; thus, this downward force causes the opposite ends of the rafters to thrust outward toward the exterior walls. Rafter thrust, as it is called, is resisted by the nailed connection of the rafter to the ceiling joists which, in turn, causes the ceiling joist to load in tension. In effect, the ceiling joist "stretches" under the load. IRC Table R802.5.1(9) specifies the minimum number of 16d nails required at each rafter-ceiling joist connection to resist these forces (see Figure 10-3).

How far a rafter can span is a function of several important factors. One of the most important considerations has to do with the amount of snow that a particular roof is expected to support. The IRC establishes the specific design criteria for different ground snow loads in Figure R301.2(5) of the code (not shown in this text). Here, expected ground snow loads for each region are provided based on historical data which, as expected, vary geographically.

Rafter spans are also a function of the particular species and grade of the lumber material. As with floor framing, wood of certain species and grades is better suited than others to support a load. The spacing of the framing members plays a role as well. Rafters and ceiling joists are commonly spaced on 24-inch (610-mm) centers, but 16-inch (406-mm) centers are also used. To a lesser degree, a spacing of 12-inch (305-mm) centers is viable. So, the snow loading criteria, species and grade of lumber, and joist spacing must be taken into account to determine how far the rafter and ceiling joists can safely span as in any other framing condition.

Fortunately, the IRC makes this easy through the use of tables such as the one in Figure 10-4. IRC Table R802.5.1(1) is printed here, which shows safe rafter spans for a variety of species and grades of wood rafters spaced at 12 inches (305 mm),

Figure 10-3

Table R802.5.1(9) Rafter/ceiling joist heel joint connections.

TABLE R802.5.1(9)
RAFTER/CEILING JOIST HEEL JOINT CONNECTIONS[a, b, c, d, e, f, h]

RAFTER SLOPE	RAFTER SPACING (inches)	20[g] 12	20[g] 20	20[g] 28	20[g] 36	30 12	30 20	30 28	30 36	50 12	50 20	50 28	50 36	70 12	70 20	70 28	70 36
		**GROUND SNOW LOAD (psf)** — **Roof span (feet)** — *Required number of 16d common nails[a,b] per heel joint splices[c,d,e,f]*															
3:12	12	4	6	8	10	4	6	8	11	5	8	12	15	6	11	15	20
	16	5	8	10	13	5	8	11	14	6	11	15	20	8	14	20	26
	24	7	11	15	19	7	11	16	21	9	16	23	30	12	21	30	39
4:12	12	3	5	6	8	3	5	6	8	4	6	9	11	5	8	12	15
	16	4	6	8	10	4	6	8	11	5	8	12	15	6	11	15	20
	24	5	8	12	15	5	9	12	16	7	12	17	22	9	16	23	29
5h12	12	3	4	5	6	3	4	5	7	3	5	7	9	4	7	9	12
	16	3	5	6	8	3	5	7	9	4	7	9	12	5	9	12	16
	24	4	7	9	12	4	7	10	13	6	10	14	18	7	13	18	23
7:12	12	3	4	4	5	3	3	4	5	3	4	5	7	3	5	7	9
	16	3	4	5	6	3	4	5	6	3	5	7	9	4	6	9	11
	24	3	5	7	9	3	5	7	9	4	7	10	13	5	9	13	17
9:12	12	3	3	3	4	3	3	3	4	3	3	4	5	3	4	5	7
	16	3	4	4	5	3	3	4	5	3	4	5	7	3	5	7	9
	24	3	4	6	7	3	4	6	7	3	6	8	10	4	7	10	13
12:12	12	3	3	3	3	3	3	3	4	3	3	3	4	3	3	4	5
	16	3	3	4	4	3	3	3	4	3	3	4	5	3	4	5	7
	24	3	4	5	5	3	3	4	6	4	4	6	8	5	6	8	10

H_c/H_R	Heel Joint Connection Adjustment Factor
1/3	1.5
1/4	1.33
1/5	1.25
1/6	1.2
1/10 or less	1.11

For SI: 1 inch = 25.4 mm, 1 foot = 304.8 mm, 1 pound per square foot = 0.0479 kPa.

a. 40d box nails shall be permitted to be substituted for 16d common nails.
b. Nailing requirements shall be permitted to be reduced 25 percent if nails are clinched.
c. Heel joint connections are not required when the ridge is supported by a load-bearing wall, header or ridge beam.
d. When intermediate support of the rafter is provided by vertical struts or purlins to a loadbearing wall, the tabulated heel joint connection requirements shall be permitted to be reduced proportionally to the reduction in span.
e. Equivalent nailing patterns are required for ceiling joist to ceiling joist lap splices.
f. When rafter ties are substituted for ceiling joists, the heel joint connection requirement shall be taken as the tabulated heel joint connection requirement for two-thirds of the actual rafter-slope.
g. Applies to roof live load of 20 psf or less.
h. Tabulated heel joint connection requirements assume that ceiling joists or rafter ties are located at the bottom of the attic space. When ceiling joists or rafter ties are located higher in the attic, heel joint connection requirements shall be increased by the following factors:

where:
H_c = Height of ceiling joists or rafter ties measured vertically above the top of the rafter support walls.
H_R = Height of roof ridge measured vertically above the top of the rafter support walls.

Figure 10-4

A portion of Table R802.5.1(1) rafter spans for common lumber species (roof live load = 20 psf, ceiling not attached to rafters).

TABLE R802.5.1(1)
RAFTER SPANS FOR COMMON LUMBER SPECIES
(Roof live load=20 psf, ceiling not attached to rafters, L/Δ = 180)

RAFTER SPACING (inches)	SPECIES AND GRADE	DEAD LOAD = 10 psf					DEAD LOAD = 20 psf				
		2 × 4	2 × 6	2 × 8	2 × 10	2 × 12	2 × 4	2 × 6	2 × 8	2 × 10	2 × 12
		Maximum rafter spans[a]									
		(feet - inches)	(feet - inches)	(feet - inches)	(feet - inches)	(feet - inches)	(feet - inches)	(feet - inches)	(feet - inches)	(feet - inches)	(feet - inches)
12	Douglas fir-larch SS	11-6	18-0	23-9	Note b	Note b	11-6	18-0	23-5	Note b	Note b
	Douglas fir-larch #1	11-1	17-4	22-5	Note b	Note b	10-6	15-4	19-5	23-9	Note b
	Douglas fir-larch #2	10-10	16-7	21-0	25-8	Note b	9-10	14-4	18-2	22-3	25-9
	Douglas fir-larch #3	8-7	12-6	15-10	19-5	22-6	7-5	10-10	13-9	16-9	19-6
	Hem-fir SS	10-10	17-0	22-5	Note b	Note b	10-10	17-0	22-5	Note b	Note b
	Hem-fir #1	10-7	16-8	21-10	Note b	Note b	10-3	14-11	18-11	23-2	Note b
	Hem-fir #2	10-1	15-11	20-8	25-3	Note b	9-8	14-2	17-11	21-11	25-5
	Hem-fir #3	8-7	12-6	15-10	19-5	22-6	7-5	10-10	13-9	16-9	19-6
	Southern pine SS	11-3	17-8	23-4	Note b	Note b	11-3	17-8	23-4	Note b	Note b
	Southern pine #1	11-1	17-4	22-11	Note b	Note b	11-1	17-3	21-9	25-10	Note b
	Southern pine #2	10-10	17-0	22-5	Note b	Note b	10-6	15-1	19-5	23-2	Note b
	Southern pine #3	9-1	13-6	17-2	20-3	24-1	7-11	11-8	14-10	17-6	20-11
	Spruce-pine-fir SS	10-7	16-8	21-11	Note b	Note b	10-7	16-8	21-9	Note b	Note b
	Spruce-pine-fir #1	10-4	16-3	21-0	25-8	Note b	9-10	14-4	18-2	22-3	25-9
	Spruce-pine-fir #2	10-4	16-3	21-0	25-8	Note b	9-10	14-4	18-2	22-3	25-9
	Spruce-pine-fir #3	8-7	12-6	15-10	19-5	22-6	7-5	10-10	13-9	16-9	19-6
16	Douglas fir-larch SS	10-5	16-4	21-7	Note b	Note b	10-5	16-0	20-3	24-9	Note b
	Douglas fir-larch #1	10-0	15-4	19-5	23-9	Note b	9-1	13-3	16-10	20-7	23-10
	Douglas fir-larch #2	9-10	14-4	18-2	22-3	25-9	8-6	12-5	15-9	19-3	22-4
	Douglas fir-larch #3	7-5	10-10	13-9	16-9	19-6	6-5	9-5	11-11	14-6	16-10
	Hem-fir SS	9-10	15-6	20-5	Note b	Note b	9-10	15-6	19-11	24-4	Note b
	Hem-fir #1	9-8	14-11	18-11	23-2	Note b	8-10	12-11	16-5	20-0	23-3
	Hem-fir #2	9-2	14-2	17-11	21-11	25-5	8-5	12-3	15-6	18-11	22-0
	Hem-fir #3	7-5	10-10	13-9	16-9	19-6	6-5	9-5	11-11	14-6	16-10
	Southern pine SS	10-3	16-1	21-2	Note b	Note b	10-3	16-1	21-2	Note b	Note b
	Southern pine #1	10-0	15-9	20-10	25-10	Note b	10-0	15-0	18-10	22-4	Note b
	Southern pine #2	9-10	15-1	19-5	23-2	Note b	9-1	13-0	16-10	20-1	23-7
	Southern pine #3	7-11	11-8	14-10	17-6	20-11	6-10	10-1	12-10	15-2	18-1
	Spruce-pine-fir SS	9-8	15-2	19-11	25-5	Note b	9-8	14-10	18-10	23-0	Note b
	Spruce-pine-fir #1	9-5	14-4	18-2	22-3	25-9	8-6	12-5	15-9	19-3	22-4
	Spruce-pine-fir #2	9-5	14-4	18-2	22-3	25-9	8-6	12-5	15-9	19-3	22-4
	Spruce-pine-fir #3	7-5	10-10	13-9	16-9	19-6	6-5	9-5	11-11	14-6	16-10
19.2	Douglas fir-larch SS	9-10	15-5	20-4	25-11	Note b	9-10	14-7	18-6	22-7	Note b
	Douglas fir-larch #1	9-5	14-0	17-9	21-8	25-2	8-4	12-2	15-4	18-9	21-9
	Douglas fir-larch #2	8-11	13-1	16-7	20-3	23-6	7-9	11-4	14-4	17-7	20-4
	Douglas fir-larch #3	6-9	9-11	12-7	15-4	17-9	5-10	8-7	10-10	13-3	15-5
	Hem-fir SS	9-3	14-7	19-2	24-6	Note b	9-3	14-4	18-2	22-3	25-9
	Hem-fir #1	9-1	13-8	17-4	21-1	24-6	8-1	11-10	15-0	18-4	21-3
	Hem-fir #2	8-8	12-11	16-4	20-0	23-2	7-8	11-2	14-2	17-4	20-1
	Hem-fir #3	6-9	9-11	12-7	15-4	17-9	5-10	8-7	10-10	13-3	15-5
	Southern pine SS	9-8	15-2	19-11	25-5	Note b	9-8	15-2	19-11	25-5	Note b
	Southern pine #1	9-5	14-10	19-7	23-7	Note b	9-3	13-8	17-2	20-5	24-4
	Southern pine #2	9-3	13-9	17-9	21-2	24-10	8-4	11-11	15-4	18-4	21-6
	Southern pine #3	7-3	10-8	13-7	16-0	19-1	6-3	9-3	11-9	13-10	16-6
	Spruce-pine-fir SS	9-1	14-3	18-9	23-11	Note b	9-1	13-7	17-2	21-0	24-4
	Spruce-pine-fir #1	8-10	13-1	16-7	20-3	23-6	7-9	11-4	14-4	17-7	20-4
	Spruce-pine-fir #2	8-10	13-1	16-7	20-3	23-6	7-9	11-4	14-4	17-7	20-4
	Spruce-pine-fir #3	6-9	9-11	12-7	15-4	17-9	5-10	8-7	10-10	13-3	15-5

(Continued)

TABLE R802.5.1(1)—continued
RAFTER SPANS FOR COMMON LUMBER SPECIES
(Roof live load=20 psf, ceiling not attached to rafters, L/Δ = 180)

RAFTER SPACING (inches)	SPECIES AND GRADE		DEAD LOAD = 10 psf					DEAD LOAD = 20 psf				
			2 × 4	2 × 6	2 × 8	2 × 10	2 × 12	2 × 4	2 × 6	2 × 8	2 × 10	2 × 12
			Maximum rafter spans[a]									
			(feet - inches)	(feet - inches)	(feet - inches)	(feet - inches)	(feet - inches)	(feet - inches)	(feet - inches)	(feet - inches)	(feet - inches)	(feet - inches)
24	Douglas fir-larch	SS	9-1	14-4	18-10	23-4	Note b	8-11	13-1	16-7	20-3	23-5
	Douglas fir-larch	#1	8-7	12-6	15-10	19-5	22-6	7-5	10-10	13-9	16-9	19-6
	Douglas fir-larch	#2	8-0	11-9	14-10	18-2	21-0	6-11	10-2	12-10	15-8	18-3
	Douglas fir-larch	#3	6-1	8-10	11-3	13-8	15-11	5-3	7-8	9-9	11-10	13-9
	Hem-fir	SS	8-7	13-6	17-10	22-9	Note b	8-7	12-10	16-3	19-10	23-0
	Hem-fir	#1	8-4	12-3	15-6	18-11	21-11	7-3	10-7	13-5	16-4	19-0
	Hem-fir	#2	7-11	11-7	14-8	17-10	20-9	6-10	10-0	12-8	15-6	17-11
	Hem-fir	#3	6-1	8-10	11-3	13-8	15-11	5-3	7-8	9-9	11-10	13-9
	Southern pine	SS	8-11	14-1	18-6	23-8	Note b	8-11	14-1	18-6	22-11	Note b
	Southern pine	#1	8-9	13-9	17-9	21-1	25-2	8-3	12-3	15-4	18-3	21-9
	Southern pine	#2	8-7	12-3	15-10	18-11	22-2	7-5	10-8	13-9	16-5	19-3
	Southern pine	#3	6-5	9-6	12-1	14-4	17-1	5-7	8-3	10-6	12-5	14-9
	Spruce-pine-fir	SS	8-5	13-3	17-5	21-8	25-2	8-4	12-2	15-4	18-9	21-9
	Spruce-pine-fir	#1	8-0	11-9	14-10	18-2	21-0	6-11	10-2	12-10	15-8	18-3
	Spruce-pine-fir	#2	8-0	11-9	14-10	18-2	21-0	6-11	10-2	12-10	15-8	18-3
	Spruce-pine-fir	#3	6-1	8-10	11-3	13-8	15-11	5-3	7-8	9-9	11-10	13-9

Check sources for availability of lumber in lengths greater than 20 feet.
For SI: 1 inch = 25.4 mm, 1 foot = 304.8 mm, 1 pound per square foot = 0.0479 kPa.
a. The tabulated rafter spans assume that ceiling joists are located at the bottom of the attic space or that some other method of resisting the outward push of the rafters on the bearing walls, such as rafter ties, is provided at that location. When ceiling joists or rafter ties are located higher in the attic space, the rafter spans shall be multiplied by the factors given below:

H_c/H_R	Rafter Span Adjustment Factor
1/3	0.67
1/4	0.76
1/5	0.83
1/6	0.90
1/7.5 or less	1.00

where:
H_C = Height of ceiling joists or rafter ties measured vertically above the top of the rafter support walls.
H_R = Height of roof ridge measured vertically above the top of the rafter support walls.
b. Span exceeds 26 feet in length.

16 inches (406 mm), 19.2 inches (488 mm), and 24 inches (610 mm) on center. Note that the heading of this table identifies that it is to be used for roof live load conditions of 20 pounds per square foot (958 Pa). Additionally, the heading states that the table is for use only where the ceiling finish material will not be installed directly to the underside of the rafters (as in a standard attic). The use of this table would not be allowed for loading conditions of greater than 20 pounds per square foot (958 Pa) or in vault ceiling conditions where the gypsum board is fastened directly to the rafter.

Enclosed attics are typically required to be ventilated per the requirements of Section R806.1. Individual rafter spaces to which ceiling coverings like gypsum board will be applied (e.g., vaulted or cathedral ceilings) must also be ventilated. Ventilated attics and rafter spaces must be provided with cross-ventilation; that is, they must be provided with a venting configuration that allows air to flow into and out of the space. Vents must be configured to prevent snow and ice from accumulating around or entering the attic or rafter space, thus obstructing air flow. Vent openings must be provided with screen mesh or similar to prevent insects from entering as well.

The amount of ventilation required is prescribed by Section R806.2. The basic requirement is that for every 150 square feet (13.94 m²) of attic space to be ventilated, 1 square foot (0.093 m²) of ventilation must be provided. Where a Class I or II vapor barrier is installed on the ceiling or where at least 50% but not more than 80% of the attic ventilation will be provided by vents in the upper portion of the attic, this quantity may be reduced by half. In the case where attic ventilation is provided by a combination of vents in the upper portion of the attic space, the balance of the required ventilation must be provided by eave or cornice vents.

Section R807.1 requires an attic access opening into all attics framed with combustible materials that are larger than 30 square feet (2.79 m²) in area and have a vertical height of 30 inches (762 mm) or more. The access is required to be located in a hallway or in some other area that is readily accessible. The minimum opening size is generally 22 inches (559 mm) by 30 inches (762 mm); however, Section M1305.1.3 (Appliances in Attics) may require a larger opening depending on the size of the largest piece of equipment installed in the attic.

GREEN OPTIONS

Like other IRC chapters that deal with framing, Chapter 8 is a great example because it allows for green options within the body of the text. In addition to the prescriptive code criteria provided for conventional lumber, the IRC also provides a number of options that utilize alternative wood products. These engineered wood products include construction materials such as end-jointed lumber, wood roof trusses, and prefabricated wood I-joists that can be used in both rafter and ceiling joist configurations.

The IRC also provides prescriptive code criteria for light-gauge steel roof and ceiling framing, just as it does with walls and floors (see Figure 10-5). Steel roof and ceiling framing offers a way to significantly reduce the wood needed to construct a dwelling because it eliminates all but the required wood roof sheathing. Steel rafters and ceiling joists form a strong, resilient roof framing system that is comparable to wood. Material standards, construction details, and span tables for steel framing are provided in the body of the code and in tables similar to those used for wood framing.

Figure 10-5

Steel C joists used to frame the ceiling and roof in a dwelling. Note the openings along the centerline of the web of the ceiling joists to accommodate wiring, pipes, etc.

COURTESY OF MARK MECKLEM, MIRANDA HOMES

Regardless of the choice in framing materials, roofs and ceilings must be structurally sound and capable of resisting all required design loads, just as floor and walls must resist all imposed loads. Roofs, ceilings, floors, and walls must form a comprehensive structural system that is capable of supporting the dwelling when exposed to all applicable loading conditions. This structural system protects not only the building's contents, but also, and more importantly, the building's occupants against the elements and forces of nature.

All of the alternative wood products explored in the following sections have been extensively tested and evaluated to ensure that they will meet all applicable code requirements and standards. The prescriptive code provisions for steel roof and ceiling framing are also based on sound engineering principles and empirical data, making them a safe and acceptable way to meet code criteria.

The IRC makes it easy to incorporate green construction into the roof and ceiling framing without having to deviate too far from basic code requirements, as one can simply utilize a green option provided in the code. In doing so, the need for expensive engineering can be partially or, in some cases, completely eliminated, because it has already been done by the manufacturers of the various engineered wood systems and those whose research culminated in the provisions for steel framing. Other options, of course, are always possible as AMM and should be explored as needed.

WOOD FRAME ROOFS AND CEILINGS

The IRC devotes a significant portion of Chapter 8 to the use of solid-sawn wood framing that is used throughout the United States and Canada. Wood roof and ceiling framing in use today encompasses much more than traditional solid-sawn lumber, however.

Wood roof and ceiling framing also includes a variety of engineered lumber products that have proven to be quite popular in recent years. In many of these cases, the IRC makes an outright allowance for these alternative wood products. Alternative products are cost effective, easy to work, and provide green options that fit within the specific materials approved for use under the code. Thus, they make excellent candidates for consideration when determining how to construct roofs and ceilings in the green dwelling project.

Section R802.1.2 End-Jointed Lumber

The IRC allows approved end-jointed lumber to be used interchangeably with solid-sawn lumber in both roof and ceiling framing. End-jointed lumber (EJL) has a number of characteristics that make it attractive for use in roofs and ceilings. It is typically straighter than sawn lumber and contains less crook, bow, and twist when compared to its traditional cousin. EJL is available in a wide variety of wood species and traditional sizes, including 2 inch by 6 inch (51 mm by 152 mm), 2 inch by 8 inch (51 mm by 203 mm), 2 inch by 10 inch (51 mm by 254 mm), and 2 inch by 12 inch (51 mm by 305 mm) nominal dimensions that are commonly used in roof-ceiling construction. Because EJL is visually graded like sawn lumber, it can be directly substituted for the traditional wood product in all roof-ceiling applications. It can also be ordered in long lengths, in some cases up to 40 feet (12,192 mm) in length, which can eliminate ceiling joist splices in many instances.

Because the IRC expressly allows EJL material to be used as a substitute for solid-sawn lumber of the same species and grade, it is relatively easy to incorporate the material into the green construction project. This interchangeability means that the prescriptive span tables provided in the code for rafter and ceiling joists can be used without modification. Thus, the same span table that works for a traditional 2 inch by 10 inch (51 mm by 254 mm) wood rafter will work for an EJL 2 inch by 10 inch (51 mm by 254 mm) of the same species and grade.

Why It's Green

End-jointed lumber is green because it is made from short segments of lumber that are fabricated into longer, usable lengths. These short segments were, at one time, considered waste products, but are now manufactured into a useful construction material. Long lengths eliminate ceiling joist splices in some cases.

Intent

The intent of this IRC section is to specifically allow the use of end-jointed lumber as a direct substitute for sawn lumber. The code further intends that, when substituting EJL for sawn lumber, consideration be given to the species and grade of the material and the specific loading conditions to which the framing member will be exposed.

Green Decisions and Limitations

Shall I incorporate end-jointed lumber into the roof or ceiling framing for my green dwelling?

How do the requirements of the IRC affect this decision?

EJL Increased Span Option

Because EJL is an engineered material, some manufacturers test their products to determine their performance and safe spans under various loading conditions. Where evaluation reports are available for EJL, increased spans may be allowed beyond those provided in the prescriptive provision of the IRC. Be sure to check these criteria closely when determining spans. EJL may actually perform better than sawn lumber in some cases, depending on the results of this testing.

Prefabricated Wood I-Joists

I-joists are also used in roofs and ceiling framing (see Figure 10-6). I-joists are relatively lightweight compared to sawn lumber, making them easy to work and handle, especially overhead. I-joists used as roof-ceiling framing are required to conform to ASTM D 5055. Most major manufacturers of wood I-joists have done extensive testing on their products and most have obtained evaluation reports

Figure 10-6

Prefabricated wood I-joists used as rafters in the roof of a dwelling. Note that conventional wood framing (rafter tails) has been added at the eaves to attach the fascia.

COURTESY OF WEYERHAEUSER

that provide the criteria for their use in roofs and ceilings. These evaluation reports, in conjunction with the manufacturer's installation instructions, provide span tables for given loading conditions (similar to those provided in the IRC for sawn lumber) as well as conditions of use, limitations, and various connection details.

I-joists can be used in virtually every roof and ceiling framing condition as sawn lumber; however, wood I-joists are not manufactured to be installed in applications where they will be exposed to the weather or in other adverse conditions. Thus, they might not be a suitable choice for, say, the exposed roof inside a swimming pool enclosure in which the framing members would be directly subjected to high humidity levels. For other unusual loading or environmental conditions, it is highly advisable to talk with the manufacturer about the specific conditions to which the wood I-joists will be exposed.

Intent

The intent of this IRC section is to specifically allow the use of prefabricated wood I-joists as an alternative to sawn lumber. The code further intends that, when substituting I-joists for sawn lumber, the specific requirements, conditions, and limitations established in the product evaluation reports and manufacturer's installation instructions must be considered.

TAKE NOTE!

Because of the "I" configuration, it is necessary to fill in the ends of an I-joist rafter and ceiling joist with wood blocking or plywood at the connection between the rafter and the ceiling joist to form a solid connection. This is necessary to ensure that the rafter thrust can be properly transferred into the ceiling joist through nailed connections. Be sure to check the manufacturer's requirements closely before splicing or connecting any engineered wood product.

Green Decisions and Limitations

Are prefabricated wood I-joists right for the roof and ceiling framing in my green dwelling project?

How does the product evaluation report impact this decision?

Increased I-Joist Depth Option

Prefabricated wood I-joists are available in a variety of joist depths. Consider using joists that are deeper than traditional sawn lumber sizes. Increased joist depths generally mean longer unsupported spans which, in turn, may mean that less additional lumber is needed for intermediate supports such as beams and girders. Additional joist depth also allows for more uncompressed insulation to be installed. Compressed batt insulation performs at significantly lower levels than uncompressed batts. So deeper joists not only reduce the amount of wood necessary for the project but also allow for improved insulation performance.

Why It's Green

Prefabricated wood I-joists are green because they use considerably less wood material than their solid-sawn cousins of comparable depth. The web material for these products is made from wood fiber that is derived from material that might otherwise be regarded as scrap material. I-joists are lighter in weight for a given length of solid-sawn material, which can potentially reduce transportation costs.

STEEL FRAME ROOFS AND CEILINGS

Roofs and ceilings framed out of light-gauge, cold-formed steel are strong and durable. Although not widely used in residential construction, steel framing is an option that is gaining popularity of late. Steel framing used in roofs and ceilings is similar in layout to that of a wood floor, in that steel joists are spaced at 12-inch (305-mm), 16-inch (406-mm), and 24-inch (610-mm) intervals. Like their wood counterparts, the allowable spans for steel framing members are also a function of the depth of the joist, the particular loading conditions, and the grade of the material used. In this case, however, the grade is determined by the particular type and strength of steel used in the manufacture of the steel joist as opposed to the species of a tree.

Steel roof and ceiling framing components must be properly identified with a mark similar to that used in wood framing. Such markings are required to identify the manufacturer of the steel joist, the uncoated steel thickness, the minimum coating designation (for corrosion resistance), and the minimum yield strength, in kips per square inch (a kip is a unit of measure used in engineering equal to a kilopound or 1,000 pounds).

It is here, however, where the similarities end. Steel roof and ceiling framing often requires special tools to cut, bore, and otherwise work the components. It is typically fastened together using self-tapping, self-drilling screws. Cutting is typically done with a steel nipper, saws, or other hydraulic tools with hardened blades and bits suitable for cutting steel. Special care must be given when cutting holes into the webs of steel rafters and ceiling joists.

Other special considerations are required for steel roof and ceiling floor framing as well. For example, where steel rafters or ceiling joists are supported by steel frame walls, each rafter or joist must be aligned directly over the supporting wall stud below. The maximum tolerance for in-line placement is 0.75 inch (19 mm) per Section R804.1.2 of the IRC.

Section R804.1 General Requirements

This IRC section charges all cold-formed steel roof and ceiling framing components to comply with the provisions of Section R804, Steel Roof Framing. This IRC section authorizes steel framing to be used in roof and ceiling framing and provides criteria by which it can be installed. Section R804.1.1, Applicability Limits, further imposes restrictions, however, on the use of steel framing under the prescriptive provisions, as the IRC does for floors and walls.

Consistent with those restrictions placed on steel frame floors, this section restricts the use of steel framing to dwellings that are not more than two stories in height; not greater than 60 feet (18,288 mm) in length measured perpendicular to the span of the joist, rafter, or truss span; and not greater than 40 feet (12,192 mm) in width measured parallel to the joist span or truss. Additionally, the IRC further restricts the use of steel framing in roofs and ceilings to sites with maximum wind speeds of 110 mph (49 m/s), exposure B or C, and snow loads not greater than 70 pounds per square foot (3.352 kPa). The IRC also imposes a range of roof slopes where steel framing can be used. For prescriptive framing, slopes must be not less than 3:12 and not greater than 12:12.

Steel load-bearing components are generally of two different configurations: C section and track section. The C-section shape receives its name from the letter shape of the joists. Track section is similar in appearance to a C section, but lacks the rolled "lip." Steel C-section joists are used in a variety of configurations, constructed to resist the loads encountered in the typical dwelling (see Figure 10-7).

In some cases, C sections are used back to back and fastened together in the web to form an I

Figure 10-7

Figure R804.3.1.1(1) Joist to rafter connection.

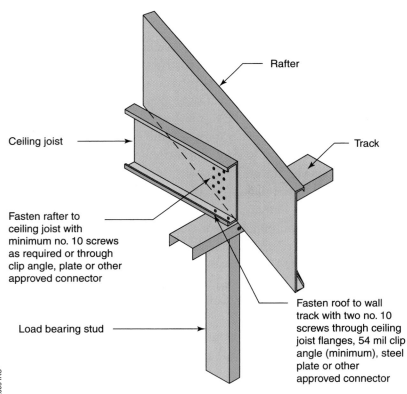

Rafter

Ceiling joist

Track

Fasten rafter to ceiling joist with minimum no. 10 screws as required or through clip angle, plate or other approved connector

Load bearing stud

Fasten roof to wall track with two no. 10 screws through ceiling joist flanges, 54 mil clip angle (minimum), steel plate or other approved connector

For SI: 1 mil = 0.0254 mm

shape, similar to that described in the preceding section. In other cases, C sections are combined with track sections to form built-up headers and girders. Again, these components are fastened together to form strong, lightweight structural supports for floor components.

The IRC provides several useful construction details that describe how to assemble various components into roof-ceiling framing members and how to frame openings. Because these details are prescriptive in nature, engineering is not required where the particular use of the steel framing is within the limits established in the code.

Intent

The intent of these IRC provisions is to provide prescriptive requirements for steel roof and ceiling framing. The code further intends that where cold-formed steel framing is used, it is used within the applicability limitation imposed by Section R804.1.1. The code further intends that steel roof-ceiling framing be capable of accommodating all imposed loads and distributing those loads to the supporting structural elements.

Why It's Green

Cold-formed steel roof-ceiling framing significantly reduces the amount of wood products needed to construct a code-compliant roof or ceiling system, and thus potentially lessens the impact to forests. Because steel framing members are not solid like sawn lumber, extra insulation can be packed or sprayed between the framing members to potentially improve thermal performance.

Green Decisions and Limitations

Shall I incorporate cold-formed steel framing into the roof and ceiling for my green dwelling project?

How do the requirements of the IRC affect this decision?

Steel Roof Truss Option

Open-web steel roof trusses can easily be incorporated into the roof design. Steel roof trusses are required to meet the American Iron and Steel Institute (AISI) Standard for Cold-Formed Steel Framing—Truss Design (COFS/Truss). Steel trusses can span considerable distances without intermediate supports and the open-web configurations make routing ducts, pipes, and wiring a relatively easy operation. They share the advantages that other steel framing components have as well. Steel roof trusses are engineered components; thus, per the requirements of Section R804.1.3, roof trusses cannot be notched, cut, or otherwise altered without specific engineering approval.

ROOF VENTILATION

The IRC requires cross-ventilation in all enclosed attics and enclosed rafter spaces where a ceiling is applied directly to the underside of the roof framing. Cross-ventilation is required to ensure the movement of air through the enclosed spaces, which aids in reducing attic temperatures during warm weather and in the control of moisture, which can migrate from the habitable portion of the dwelling through ceilings and into the attic space. Openings into the attic space or rafter cavities used for ventilation are regulated under the provisions of Section R806.1 to ensure they are large enough to allow for the passage of air but small enough to prevent insects from entering.

Minimum required ventilation areas are defined in IRC Section 806.2 to ensure that an adequate volume of air from outside the dwelling can be delivered to the enclosed attic or rafter space. The introduction of outdoor air for ventilation is not without its share of difficulties, however. In the depths of summer or winter, air temperatures within enclosed ventilated attic spaces can be extreme, a factor of major influence on heating and cooling bills. This is especially true where air ducts and water supply piping are located within the attic space and subjected to these extremes of temperature. Fortunately, green options exist for attics just as they do in other areas of the code.

Section 806.4 Unvented Attic Assemblies

A relatively new option exists in the IRC for the construction of attics. Beginning with the 2006 edition of the IRC, the code allows "unvented attic assemblies" (the space between the ceiling joists of the top story and the roof rafters) to be constructed per the conditions outlined in this section.

An unvented attic assembly, as the name implies, is an attic space where the ventilation normally required by Section R806.1 is omitted and various combinations of air-impermeable and/or air-permeable insulation are applied directly to the underside or the "interior" side of the structural roof deck (see Figure 10-8). Furthermore, the insulation and vapor barrier that would normally be provided on the flat ceiling

Figure 10-8

Spray-applied foam insulation installed in an unvented attic. Note that the insulation is applied directly to the roof deck and gable-end wall beyond.

COURTESY DELMAR/CENGAGE LEARNING

are eliminated as well. In this way, the attic space becomes, essentially, part of the conditioned space within the dwelling. This method of roof-ceiling construction effectively "tempers" the attic space and prevents large temperature fluctuations as would otherwise be the case in a traditional ventilated attic.

The primary advantage of such a space is it creates a more moderate environment in which to locate HVAC equipment and duct work. The range of temperatures within these unvented attic assemblies is *relatively* constant, because the temperatures within these spaces tend to not fluctuate to the extreme as would otherwise be the case in a ventilated attic. Because temperatures within these spaces are more moderate, equipment and especially duct work can perform much more efficiently than if they were installed in an attic where temperatures may fluctuate to great extremes.

Moisture Control in Unvented Attic Assemblies

When warm, moist air generated within a dwelling (e.g., from bathing and cooking) comes into contact with a cold surface (such as in a ventilated attic), moisture can condense and accumulate on attic surfaces. In small quantities, this moisture can be harmless. Where excessive amounts are generated or are present for long periods of time, however, excess moisture can result in mold growth or decay, which is undesirable by any standard.

It is important to note that, while the use of the unvented attic option allowed in IRC Section R806.4 creates the notion that such an attic is indeed a "conditioned" space, it is not regarded as a *true* conditioned space like the habitable portion of the dwelling contained within the heated envelope. Instead, according to the 2006 IRC Code and Commentary Volume 1, it recognizes that the attic area is "indirectly conditioned because of omission of the air barrier, insulation at the ceiling and leakage around attic access opening." The IRC does not require the unvented attic to be provided with a source of conditioned air (heating or cooling) to maintain the required temperature as it does with the habitable portions of the house; therefore, there is no assurance that attic temperatures will be maintained in the same way as within the occupied portion of the dwelling.

Thus, moisture control is critical in unvented attics, as it is in ventilated attics, to avoid condensation. In some cases, this problem is controlled by virtue of the fact that the unvented attic space is located in a climate zone where moisture and condensation is not a problem. In other cases, the problem is controlled because the unvented attic assembly is indirectly conditioned and is thus tempered to a useful degree without the need for special construction.

In still other cases, however, extra precautions must be taken to ensure that moisture will not be a problem. IRC Section R806.4, Condition 5, establishes three cases where special steps must be taken to ensure that the unvented attic assembly will not be subject to moisture problems. The first case requires the use of air-impermeable insulation exclusively against the roof sheathing, which effectively isolates the moist air from the cold underside of the roof deck by acting as a vapor retarder.

In the second case, where air-permeable insulation is used exclusively, the IRC requires the air-permeable insulation to be applied to the underside of the roof sheathing but imposes the additional requirement to provide rigid board or sheet insulation to be installed on top of the structural roof deck in the R-values identified in Table R806.4 for condensation control.

Why It's Green

Unvented attic assemblies are green because they create a large, tempered space above the habitable area of a dwelling. This space can eliminate extremes of temperature that might otherwise occur in a ventilated attic, creating an excellent environment for HVAC equipment and duct work. Efficiency of these components is greatly improved when installed in unvented attics which, in turn, can mean less energy usage and enhanced thermal performance overall.

In the third case, where both air-impermeable and air-permeable insulation types are combined, IRC requires the air-impermeable insulation to be installed in direct contact with the underside of the roof sheathing and the air-permeable insulation to be installed directly under the air-impermeable insulation. In these ways, temperatures are maintained at a sufficiently high temperature to avoid condensation within the unvented attic assembly.

Intent

The intent of this IRC section is to allow an unvented attic alternative to the traditional ventilated attic. The code further intends that careful consideration be given to climatic conditions when determining the thicknesses of insulation required and also that consideration be given to vapor barrier requirements for condensation control. The code also intends that air-impermeable insulation be tested in accordance with either ASTM Standard E 2178 or Standard E 283.

Green Decisions and Limitations

Shall I incorporate an unvented attic assembly into the design of my green dwelling?

How will the requirements of the IRC impact this decision?

HVAC Location Option

Where unvented attics are a viable option, locate HVAC equipment and ducts within this space when possible (see Figure 10-9).

Figure 10-9

Duct work installed in an unvented attic.

COURTESY OF DEMILEC (USA) LLC®

TAKE NOTE!

Spray-applied foam insulation products that are typically used in unvented conditioned attics are regarded as foam plastics under the IRC and are thus subject to the requirements of Sections R316 Foam Plastic and R302.10 Flame Spread Index and Smoke Developed Index for Insulation. Some of these products may require a thermal barrier, an ignition barrier, or both, depending on circumstances. Many of these insulation products have been extensively tested by manufacturers and have evaluation reports available regarding their use. Some evaluation reports establish additional requirements depending on specific installation conditions. Be sure to read and thoroughly understand all code and manufacturer's criteria when selecting a product for use.

Ducts routed through these attic spaces perform better than those that run through traditional ventilated attics, because this practice eliminates the need to run heated air through very cold spaces in winter and cooled air through very warm spaces in the summer. Improved duct performance can mean less heating or cooling loss, which potentially means lower energy bills. Conditioned attic spaces are more comfortable, thus service work can be more easily accomplished. Pipes that run through unvented conditioned attics are provided with additional freeze protection as well, and benefit from the same improved energy efficiencies as ducts.

CHAPTER 10 APPLIED TO THE SAMPLE PROJECT

Chapter 10 pertains to roof and ceiling framing systems of sawn wood, steel framing, and engineered wood systems. Additionally, Chapter 10 addresses attic access, attic ventilation, and the treatment of unvented attic assemblies. The following analysis will demonstrate the application of some IRC code provisions for roofs and ceilings with consideration given for the proposed scope of work and our design philosophy.

Site Considerations

The fire-resistive construction requirements of Section R302.1 are imposed due to the placement of the exterior wall at 3 feet (914 mm) from the property line as we have already seen. Assuming that the roof of our sample dwelling has an eave projection of only 1 foot (305 mm) beyond this wall, the edge of the eave and roof assembly will be located only 2 feet (610 mm) from the property line. Table R302.1 requires the underside of projections (such as the eave of our sample dwelling), located less than 3 feet (914 mm) from a property line, to be protected with 1-hour fire-resistive construction.

Building Considerations

The roof and ceiling will be framed with light-gauge steel framing, which is consistent with our desire to minimize the use of wood wherever possible, just as it was with the interior wall framing. Section R804 of the IRC already addresses steel frame roofs and ceilings so it will be relatively easy to incorporate steel into the sample project. The necessary steel thicknesses, connection details, and specifications are outlined in this section of the code.

Section R804 allows the use of prescriptive steel framing, outlined as a preference in our design philosophy. Prescriptive framing typically eliminates the need for expensive engineering; however, because we will need the services of a design professional for the design of the rammed-earth exterior walls, we will also want to discuss the use of prescriptive steel framing in the project. Many engineers will prefer and may even be required to design the entire structure to ensure that all loads are considered and all code requirements are met, with the added benefit that she may be able to value-engineer the project to our advantage.

Additionally, we outlined in the scope of work that we would incorporate the unvented attic option allowed by Section R806.4 and also that we would incorporate spray-applied foam insulation. Section R806.4 establishes many requirements for this option so we will need to read these provisions carefully to ensure compliance. Also, depending on the climate zone in which our sample project will be built, IRC Table R806.4 may require additional insulation for condensation control. We may also have to add insulation to account for the increased thermal conductivity of the steel framing, yet another discussion point for our meeting with the code official.

Finally, because IRC Section R302.1 requires the underside of the eave to be protected with 1-hour fire-resistive construction, we will want to provide a construction detail in our design that ensures this requirement will be met. Let us also add this item to our growing list of construction topics to discuss with the code official as she may be able to advise us how to best accomplish this.

CHAPTER SUMMARY

Chapter 8 of the IRC addresses the various code requirements for roof-ceiling framing systems. It provides standards for construction materials and methods of installation for both wood and steel construction materials. Many of the materials covered in the chapter offer green alternatives that are accepted under the provisions of the code. Following are highlights of the chapter.

- Chapter 8 of the IRC provides prescriptive code requirements for both wood and steel frame roofs and ceilings.
- Criteria for roof and ceiling framing systems are provided in both text and tabulated form.
- Span tables are provided for both wood and steel construction.
- Wood and steel roof-ceiling framing systems collect and distribute loads to supporting elements such as walls and beams.
- A wood or steel rafter or ceiling joist's allowable span is a function of several criteria including required load, spacing of the member, the species and grade of the material if wood or the grade of the steel, and allowable deflection criteria, as it is for floor framing.
- Many of the applications available for roof and ceiling framing are already "green" alternatives and can simply be chosen as an allowed option.
- The manufacturer's installation guidelines for engineered wood products must be observed.
- Unvented attic assemblies provide a highly efficient means to insulate an attic space.
- The performance of HVAC equipment and duct work can be improved if installed within an unvented attic space.
- Moisture must be carefully controlled in unvented attics.

ROOF COVERINGS

This chapter corresponds to IRC Chapter 9, Roof Assemblies

IRC CHAPTER 9 - OVERVIEW

Chapter 9 of the International Residential Code provides requirements for roof assemblies or "coverings" as they are more commonly known. The chapter begins with Section R901.1, Scope, which requires the design, materials, and construction of all roof assemblies to comply with the provisions of the chapter. This charging statement is important because it establishes that the provisions of the chapter apply to *all* buildings that fall under the scope of the IRC. So, this chapter applies as much to townhouses and dwellings as it does to accessory buildings and garages.

Roof coverings have two primary duties. First, they must provide weather protection for the building. Section R903.1 requires the decks for all roofs to be covered with *approved* roof coverings that have been properly secured to the building. The term "approved," as it is defined in the code, means acceptable to the building official. Roof coverings must be deemed suitable by the building official before they can be used. For traditional materials this is relatively easy, as most of these roof coverings are already addressed in the code. For those roof coverings that are not addressed in the code, they are treated as AMM as we have seen previously.

Roof coverings, then, must first be found suitable as a weather barrier. Section R903.1 also requires the materials to be properly fastened to the building. This is to ensure that roof coverings stay securely in place during extremely adverse conditions, an important consideration in areas with high winds or other severe weather conditions. The IRC prescribes the method of attachment for most commonly installed roof coverings (see Figure 11-1). Additionally, the IRC requires roof assemblies to meet all of the requirements of the code in addition to the

Figure 11-1

A typical composition roof covered with architectural grade shingles. Note the use of asphalt felt paper and metal flashing at roof-wall intersections to prevent leaking.

COURTESY DELMAR/CENGAGE LEARNING

manufacturer's installation instructions, as it does for most other manufactured building products.

IRC Section 903.2 requires flashing to be used at all roof-wall intersections, at copings, at moisture-permeable building materials, and any other location where water might infiltrate into the building. Other locations typically include changes in the direction of roof slope and around openings into the building. Where metal flashing is used, it must be corrosion resistant and a minimum thickness of No. 26 galvanized sheet. Where parapet walls are encountered, Section 903.3 requires coping of noncombustible, weatherproof materials to be used.

Roofs must not only resist the infiltration of water into the building, but also properly drain water off of and away from the structure. Roofs are typically sloped to some degree to allow gravity to move water to and off of the edge of the roof. The degree of slope varies depending on the type of roof covering material chosen, in addition to aesthetic considerations. Some roofs are "flat" (barely sloped) or do not readily allow water to drain off of the edge of the roof, such as when parapet walls are provided. In these cases, roof drains are required (see Figure 11-2). Roof drainage can be provided by traditional drains or **scuppers** (openings in the parapet wall that allow drainage). Where roof drains are required, so are overflow drains, as a backup drainage system should the primary system become blocked. Where installed, roof drains, overflow drains, leaders, and conductors must meet the requirements of the International Plumbing Code (IPC) or another plumbing code adopted by the local jurisdiction.

The second important consideration for roof coverings is they must meet the roof classification requirements of Section R902 of the IRC. Section R902.1 requires all roofs to be covered with materials as set forth in the code and, more important, to be covered with roof materials meeting Class A, B, or C designations in locations that are required by law and in those cases where the edge of the roof is located less than 3 feet (914 mm) from the property line. Roof materials must be tested in accordance with UL 790 or ASTM E 108 standards. Roofs covered with slate, concrete tile, or other traditional, noncombustible materials are deemed to meet the requirements for Class A roof coverings.

So, why does the IRC require roofs to be Class A, B, or C rated? It does so largely to prevent conflagration by spread of fire from one roof to the next during fire conditions. During a fire, the wind can carry burning debris from one building

Figure 11-2

A new roof drain installed during a reroof, prior to the installation of the roof covering materials. Note the steel seismic strap ties that were retrofit as part of a structural upgrade.

COURTESY OF FROET INDUSTRIES, LLC

remember

In areas where buildings are built less than 5 feet (1,524 mm) from a property line, the IRC requires exterior walls to be fire-resistance rated for a period of not less than 1 hour. Where walls are less than 3 feet (914 mm) from a property line, openings are prohibited to prevent the spread of fire from one building to the next through windows, doors, and penetrations into the wall. These protections also extend to the roof assembly as described in the previous sections.

to another and deposit it on the roof. Roofing materials with particular class ratings provide some resistance to ignition when this occurs. In other areas, local laws might require certain class ratings for similar reasons. For example, local laws might require roofs for dwellings and accessory buildings to meet the Class A (essentially, noncombustible) designation in areas prone to wildfires so that a minimum degree of fire protection is provided where burning embers might be deposited on the roof in fire conditions.

For many of the traditional roof covering materials, underlayment materials, and weather-resistive barriers addressed in the code, ASTM standards are referenced. In addition, standards from the Cedar Shake and Shingle Bureau (CSSB)

Figure 11-3

Table R905.7.5 Wood shingle weather exposure and roof slope.

TABLE R905.7.5
WOOD SHINGLE WEATHER EXPOSURE AND ROOF SLOPE

ROOFING MATERIAL	LENGTH (inches)	GRADE	EXPOSURE (inches)	
			3:12 pitch to < 4:12	4:12 pitch or steeper
Shingles of naturally durable wood	16	No. 1	3¾	5
		No. 2	3½	4
		No. 3	3	3½
	18	No. 1	4¼	5½
		No. 2	4	4½
		No. 3	3½	4
	24	No. 1	5¾	7½
		No. 2	5½	6½
		No. 3	5	5½

For SI: 1 inch = 25.4 mm.

for wood shake and shingle installations are referenced, as are other standards in the International Building Code.

The IRC provides criteria for the method of attachment for the various products used to cover roofs. Specific sizes, types, and spacings of fasteners are prescribed for each of the roof coverings addressed in the code, as are weather exposure and lap requirements. The attachment and weather exposure criteria are provided in either text form or tables, such as Table R905.7.5 shown in Figure 11-3. This table prescribes the weather exposure and roof slope criteria for wood shingle roof coverings, a material that is naturally resistant to decay.

GREEN OPTIONS

As we saw in Chapter 7 of the IRC, which dealt with wall coverings, the green options available to an individual considering what sorts of materials to use as roof coverings are largely dependent on how one defines "green" in the first place. This is just as true for Chapter 9 of the IRC as it is for other chapters under our review. What direction an individual takes in the decision to use this green method or that is, again, largely one of personal preference.

Whether one's perspective of green means the use of natural materials for roof coverings, such as wood products, or man-made materials, such as metal sheeting, there are many green choices available. In the case of natural materials, wood shakes or shingles might be the appropriate choice, and the IRC provides criteria specifically for those materials including lap and fastener requirements. In the case of man-made materials, metal roofing or fiberglass shingles might be the appropriate choices, and again the IRC provides similar criteria for these materials. For some, going green is based on a combination of factors to be carefully considered.

Given that so many variables affect the decisions to be made in choosing materials for roof coverings, this text makes no attempt to prescribe which of these options is the most green. The reader decides which of these green options is the most meaningful and then chooses a course of action based on the requirements of the IRC and the suggestions provided here.

ROOF COVERINGS OF NATURAL MATERIALS

As the name implies, roof coverings derived from "natural" materials are those made from products taken directly from nature, such as wood and slate. Wood shakes and shingles are natural wood products split or cut from cedar logs. Slate tiles are split from slate slabs quarried from the earth. These products are time-tested roof covering materials and provide an effective barrier against the elements when properly installed. Additionally, they are recognized by the IRC as approved materials. Methods of installation are spelled out in the code prescriptively; thus, it is easy to incorporate them into the green project simply by following the code's requirements.

As with most roof coverings, those derived from natural materials require regular maintenance, although slate requires less than wood. Due consideration should be given to this fact when considering the use of these materials in a dwelling. If you are considering the use of an AMM for a roof covering, it will need to meet the general requirements outlined in the code for conventional roofs, meet the intent of the IRC, and be deemed to be at least the equivalent of those conventional materials specified in the code.

Section R905.6 Slate and Slate-Type Shingles

Slate roofs have been around for centuries. They create a stately, elegant appearance and are inherently durable, which makes it easy to understand why they are prized in residential construction. Slate is similar to shale in its composition of thin layers of rock that are relatively easy to split. Slate shingles are made from larger slabs of slate by splitting the layers until the shingles are the desired thickness (see Figure 11-4). They are then trimmed or cut into the desired size and shape for use as a roofing material. Slate shingles must comply with ASTM C 406.

Figure 11-4

A genuine slate roof. Note the random mix of grey, green, and plum colored slate tiles.

Slate is a fairly heavy roof covering material, weighing 7 to 8 pounds per square foot (335 to 383 Pa) depending on shingle thickness. The lightest asphalt shingles, by comparison, weigh 2.25 pounds per square foot (108 Pa). Thus, the additional weight of the slate roof covering must be taken into account when sizing rafters and other supporting elements. When using the prescriptive rafter span tables found in Chapter 8 of the IRC with slate roofs, be sure to use the columns with dead loads that are representative of the actual assembly of the roof including the slate tile.

IRC Section R905.6.1 requires slate to be installed on solid sheathing such as plywood or OSB. Other IRC sections require a minimum roof slope of 4:12 or greater where slate shingles are used and a minimum **headlap** (shingle overlap) of 2 inches (51 mm) to 4 inches (102 mm), depending on roof slope as prescribed by IRC Table R905.6.5. A minimum of two fasteners are required. The IRC specifies special flashing and joint treatment for slate tile to ensure that seams and intersections will not leak.

Intent

The intent of this IRC section is to establish minimum materials and installation standards for the use of slate and slate-type roofing products in dwellings. The code further intends that materials used in the application of slate roof coverings meet ASTM 406 standards for quality. This code section also intends that slate roof shingles be securely attached to the structure and that they be provided with an underlayment meeting the requirements of ASTM D 226, Type I or ASTM D 4869, Type I or II.

Green Decisions and Limitations

Shall I use a green alternative to traditional slate roof shingles on my green dwelling?

How will the requirements of the IRC impact my decision?

Options

When selecting roof covering materials that will meet the requirements of IRC Section R905.6, consider the following green alternatives.

Reclaimed slate: Because slate shingles are derived from a naturally durable rock product, they enjoy a useful life measured in decades, which can be considerably longer than the life of the original structure on which they were placed. Reclaimed slate shingles are gathered from deconstruction projects

Why It's Green

Alternatives to traditional slate, such as reclaimed slate or synthetic slate, are green because they utilize materials that have either already been taken from nature (as with reclaimed slate) or are manufactured from materials that do not require quarrying (as with synthetic slate). The use of recycled and reused materials is generally regarded as a sustainable practice. Synthetic slate materials are man-made and can provide an attractive alternative to traditional slate where it is unavailable regionally or where there is a desire to avoid the use of materials that must be removed from nature.

Figure 11-5

Reclaimed slate shingles. Note the coloring, mineral deposits, and textures due to weathering.

COURTESY OF ISTOCKPHOTO.COM

TAKE NOTE!

The use of reclaimed or "used" construction materials requires the approval of the code official as discussed in Chapter 5 of this text. Be sure to check with the code official *before* you purchase, install, or otherwise make a commitment to use reclaimed materials in your green dwelling project.

and then sold through salvage companies. Reclaimed slate shingles have a charming, weathered appearance that can make them attractive for architectural reasons; however, they can be fragile and must be removed carefully to avoid damaging them (see Figure 11-5). Reclaimed slate can also be obtained where the building owner simply wants to employ a different look and chooses to replace them with other materials. Either way, the practice of reusing construction materials makes good sense and is inherently sustainable.

Synthetic slate roofing: Although synthetic or man-made construction materials have been around for many years, recent advancements in technology have allowed extremely convincing synthetic products to be developed. Synthetic slate products are no exception, as shown in Figure 11-6. Because the appearance of synthetic construction materials has improved, so has their popularity

Figure 11-6

A roof covered with a synthetic slate roofing material.

increased in recent years. Synthetic slate products include individual tiles, which are similar in appearance to real slate shingles, and three-tab versions, similar to fiberglass or composition shingles. Some products are made from 100% recycled materials (e.g., used tires), making them even more attractive as a green construction material. These products must be installed per manufacturer's specifications. Also, check with the local building official when considering the use of man-made slate materials to ensure that the specifications for the proposed material are acceptable to him.

Section R905.7 Wood Shingles and Section R905.8 Wood Shakes

Like slate, wood shingles and shakes have been used for centuries as a roof covering (see Figure 11-7). What is the difference between a wood shingle and a shake? Wood shingles are milled and made smooth on both sides to create a neat, tidy appearance. Shingles are typically cut on a bevel, with one end thicker than the other. Shakes, in contrast, are split from wood billets and are left with a rougher, more rustic looking, irregular surface. Cedar is commonly used for shingles and shakes because of its natural durability.

Shingles and shakes may be installed on solid or spaced sheathing as specified in IRC Sections R905.7.1 and R905.8.1. Where spaced or "skip" sheathing is used, it must be spaced at a dimension that matches the weather exposure of the shingle or shake to ensure that fasteners will penetrate into the wood. Roofing underlayment must comply with ASTM D 226, Type I or ASTM D 4869, Type I or II. Wood shingles and shakes must be of a suitable grade as specified

Figure 11-7

Natural wood shakes installed on a roof.

by the CSSB. Grades for each type of roof covering vary based on material used and application.

Wood shingles and shakes are required to be installed with both a minimum side lap and a minimum headlap, which again varies by application. Fasteners for use in the attachment of wood shingles and shakes must be corrosion resistant and must penetrate into the roof sheathing a minimum of 0.5 inch (13 mm). Where the roof sheathing is less than 0.5 inch (13 mm) thick, fasteners must penetrate fully through the sheathing.

Due to the surface characteristics of wood shakes, it is necessary to interlay No. 30 roofing felt in 18-inch (457-mm) wide strips between each course of shakes as required by IRC Section R905.8.7. The felt must be installed in such a way that no portion of it is exposed to the weather. Both wood shingles and shakes must be properly flashed at roof valleys, wall intersections, and openings into the roof to ensure that water will not seep through the roof.

Intent

The intent of this IRC section is to establish minimum materials and installation standards for the use of wood shingle and shake roofing products in dwellings. The code further intends that materials used in the application of wood roof coverings meet CSSB standards for quality. This code section also intends that wood roof shingles and shakes be securely attached to the structure and that they be provided with an underlayment meeting the requirements of ASTM D 226, Type I or ASTM D 4869, Type I or II.

Green Decisions and Limitations

Shall I use preservative-treated wood or synthetic wood products in my green dwelling?

How will the requirements of the IRC impact this decision?

Options

When selecting roof covering materials that will meet the requirements of IRC Sections R905.7 and R905.8, consider the following green alternatives.

Preservative-treated shingles and shakes: Wood preservatives have been used to extend the life of wood construction materials since the early 1900s. The use of preservative treatments for wood products can greatly enhance the service life of wood products, in particular where the wood products will be exposed to extreme weather conditions such as on a roof (see Figure 11-8). A prolonged service life means that less maintenance is likely to be required overall and that the replacement of the roof coverings should not be needed as frequently as would otherwise be the case. Preservative-treated wood shingles and shakes are installed in the same manner as nontreated shingles and shakes.

> **Why It's Green**
>
> Preservative-treated wood products are treated with a chemical that greatly improves the life of the wood. A longer service life potentially means less maintenance and replacement costs for the wood roof, which is good for forests. Synthetic wood products are man-made and do not utilize wood products in their manufacture. This potentially reduces the impact to forests by decreasing demand for wood products.

Figure 11-8

A preservative-treated roof shingle. Note the beading of water on the surface from the preservative treatment, which increases the life of the shingle considerably.

COURTESY OF CEDAREX™ — WWW.CEDAREX.COM

It is important to use suitable fasteners for the preservative-treated wood. The IRC generally requires fasteners for all roof covering materials to be corrosion resistant to ensure durability and to avoid streaking and staining that may occur when other fasteners rust. Some preservative materials may react with metal fasteners so it is important to use those that are suitable for the preservative treatment.

Synthetic wood shingles and shakes: Like synthetic slate roofing products, the appearance of man-made roofing shingles and shakes has greatly improved in recent years. Many such products are virtually indistinguishable from real wood at a distance (see Figure 11-9). These products are lightweight, durable, and easy to install. Many of these products have evaluation reports available. In such cases, follow the recommendations of the manufacturer

Figure 11-9

A realistic-looking synthetic wood shingle material.

COURTESY OF DAVINCI ROOFSCAPES

and the evaluation report closely. Synthetic wood roof coverings reduce or eliminate the need for wood products, which can contribute to the improved health of forests. Many of these products are made in part or entirely from recycled materials, an obvious choice for use in a green dwelling!

MAN-MADE ROOF COVERINGS

As we have seen already, man-made roof covering materials can be manufactured to closely resemble natural products. Some man-made materials, however, make no attempt to mimic other products because of their unique charm. For example, clay roof tiles are used extensively in the western United States and have a unique character all their own. And, let us not forget ecoroofs, which were studied at length in Chapter 3 of this text.

Clay roof tiles, concrete tiles, and similar roof covering materials are manufactured from products found in nature, from man-made materials such as concrete, or in various combinations of the two. In any case, these products are also highly durable and greatly minimize the amount of maintenance necessary to keep them looking good. Man-made roof covering products also include asphalt, fiberglass, and composition roof shingles and metal roofs.

The IRC prescribes methods of attachment for many common forms of man-made roof coverings. Depending on the type of material specified, the text of the code provides attachment criteria for quantity and length of fasteners and other relevant criteria. Additionally, the IRC requires close adherence to the manufacturer's installation criteria.

Figure 11-10

Composition roofing installed on a dwelling.

COURTESY DELMAR/CENGAGE LEARNING

Section R905.2 Asphalt Shingles

This IRC section provides code requirements for the installation of asphalt shingles used as roof coverings in dwellings. Broadly, this category also includes fiberglass and composition shingles. Asphalt shingles (shingles, hereafter) and their related cousins come in individual shingle forms and three-tab forms (see Figure 11-10). These shingle products come in a variety of colors, textures, and styles and are commonly used in all areas of the United States.

Some shingles are called "architectural" and are manufactured in thicknesses greater than that typically found with standard three-tab shingles. Some of the architectural products come in varying thicknesses, lengths, and textures so as to create a roof with more visual interest than standard shingles. Colors available in architectural grade shingles also vary but in many cases are manufactured to more closely resemble natural products in earth tones. The installation of all such products is similar; however, as with all manufactured construction materials, the installation criteria from both the code and the manufacturer must be followed.

Shingles of all types must be installed on solid sheathing and, as we have seen with other roof coverings, the underlayment used beneath the shingles must meet ASTM D 226, Type I requirements. Additionally, the IRC allows ASTM D 4869, Type I or ASTM D 6757 standards to be used for underlayment as well. Asphalt shingles must conform to the wind loading requirements established in the IRC and must be tested per the terms of ASTM D 7158. For asphalt shingles not included in the scope of ASTM D 7158, these must be tested and labeled to indicate compliance to ASTM D 3161. The standard ASTM D 3161, Class F is suitable for all areas.

Asphalt shingles must not be installed on slopes less than 2:12. For roof slopes between 2:12 and 4:12, double underlayment is required per Section R905.2.2. Fasteners for all shingle applications must be galvanized steel, stainless steel, or other approved roofing nails and must be long enough to penetrate into roof sheathing at least 0.75 inch (19 mm). Shank diameters for roofing nails must be at least 12 gauge (3 mm) with a head diameter of not less than 0.38 inch (10 mm) per ASTM F 1667, to ensure that the shingle will stay attached to the roof deck under extreme wind conditions.

As with all roof coverings, shingles must be properly flashed at all roof-wall intersections, at all openings into roofs, and at changes in direction such as occur at valleys. Section R905.2.8 provides requirements here. Additionally, the IRC provides criteria for the treatment of saddles and crickets where roof slopes intersect surfaces on the downward side and parallel to the roof slope, such as occurs at chimneys. This is to ensure that water adequately drains around the obstruction.

Intent

The intent of this IRC section is to ensure that shingles used as roof coverings are suitable for use and are installed to widely recognized standards. The code further intends that roof coverings be properly attached to the supporting structure. The IRC also intends that roofs be properly flashed to prevent the infiltration of water at changes in direction, slope, and at obstructions.

Green Decisions and Limitations

Are there green alternatives to traditional asphalt shingles for use as roof coverings?

How does the IRC impact this decision?

Why It's Green

Some green options include the use of recycled materials in the manufacture of the roof covering. Using recycled content to make a product is considered by many to be a sustainable practice. Other green options include using nonwood materials, to reduce demand for forest products and potentially improve the health of forests.

Options

When selecting roof covering materials that will meet the requirements of IRC Section R905.2, consider the following green alternatives.

Shingles made from recycled materials: Some manufacturers are now using recycled plastics from the waste products of plastic grocery bags, used carpeting, and recycled tires in the manufacture of roof coverings. Others use a combination of recycled plastic and wood fiber as is done with some siding materials. Shingles made from these products are as durable as those made in the traditional way, but utilize materials that have already served a useful life and would otherwise end up in a landfill. Consider using roof covering materials that contain at least some recycled content (see Figure 11-11). Better yet, consider the use of materials made entirely of recycled materials.

Metal shingles and sheet roofing: Some man-made roof coverings utilize metal in the form of individual shingles or metal sheets. These strong, lightweight materials are allowed under the provisions of the IRC. Metal roof assemblies use no wood products

Figure 11-11

A man-made shake roof material made from 75% recycled content. In this case, the material used is recycled tires.

COURTESY OF EUROSHIELD

Figure 11-12

Stone-coated metal roofing tiles.

COURTESY OF METRO ROOF PRODUCTS

and, in most cases, require little if any maintenance. Metal roof shingles can be installed on roof slopes greater than 3:12 and come in a variety of styles and colors (see Figure 11-12). Similarly, metal sheet roofs come in a variety of styles and colors but some types can be installed on slopes as flat as 0.25 unit vertical to

TAKE NOTE!

Metal roofs may generally be installed on either solid or spaced roof decks; however, some metal roof assemblies are manufactured to be installed on spaced sheathing. In these cases, it is mandatory that spaced sheathing be installed per Sections R905.4.1 and R905.10.1.

12 units horizontal. It is imperative that manufacturer's criteria be followed when selecting and installing metal sheet roof coverings. Some metal roofs are made from recycled materials and can be easily recycled when they have served their useful life.

CHAPTER 11 APPLIED TO THE SAMPLE PROJECT

Chapter 11 pertains to roof assemblies or, as they are more commonly called, roof covering materials. Chapter 11 addresses the use of roof coverings including slate, wood shakes and shingles, synthetic shake and slate materials, and asphalt shingles. This chapter also addresses requirements for roof classification for fire resistivity, underlayment, fasteners, standards for various roof coverings, and specific conditions affecting the installation and use of various roof assemblies. The following analysis will demonstrate the application of some IRC code provisions for roof coverings with consideration given for the proposed scope of work and our design philosophy.

Site Considerations

As noted in the previous chapter, the eave projection for the sample project was assumed to extend 12 inches (305 mm) beyond the edge of the fire-resistive wall. Thus, the edge of the roof assembly will be located only 2 feet (610 mm) from the property line, triggering the requirements of Section R902.1 of the IRC which requires the use of Class A, B, or C roofing material where the edge of the roof is located less than 3 feet (914 mm) from a property line.

Building Considerations

We must carefully select a roofing type that will satisfy the requirements of both the code and our design philosophy. It is clear that, whatever product we choose, we must satisfy the IRC requirement to use a Class A, B, or C roofing product as roof covering for the project. Recall that we also identified the desire to minimize the use of wood products and also to utilize recyclable (or, presumably, products *made* from recycled materials as well) to the maximum degree possible in our design philosophy. Thus, it seems that a synthetic slate tile or synthetic wood shingle material made at least in part from recycled materials will work well for our roof covering materials, assuming we can find a product with the appropriate classification.

As we do our research, we must ensure that the synthetic slate or shingle roofing we choose also meets either standard ASTM D 225 or ASTM D 3462 as is required for all shingles, and wind resistance standard ASTM D 7158 or ASTM D 3161, depending on regional wind speed requirements, as the case may be. Additionally, we must be sure that our chosen roof covering material is made from recycled materials and/or can be recycled at the end of its useful life to satisfy our self-imposed design conditions.

CHAPTER SUMMARY

Chapter 9 of the IRC addresses roof coverings. It provides standards for materials and methods of installation for a variety of commonly used construction materials. Many of the materials covered in the chapter offer green alternatives to conventional options, or are regarded as green themselves. Following are highlights of the chapter.

- The IRC regulates roof assemblies (coverings) for all habitable and accessory buildings that fall under the scope of the code.
- Criteria are provided in the text of the code as well as in tabulated form. Methods of attachment, weather exposure criteria and type, size, and spacing of fasteners are identified.
- Roof assemblies must meet the weather protection requirements of the IRC.
- Roof assemblies must meet the roof classification requirements of the IRC.
- Roof assemblies must be properly flashed at all roof-wall intersections, changes in roof slope or direction, and at openings into the roof.
- Roof assemblies must be provided with adequate drainage. Where roof drains are required, they must meet the requirements of the IPC.
- Manufacturer's installation criteria must be followed as directed by the IRC to ensure a code-compliant installation and to preserve the manufacturer's warranty.
- Some traditional roof covering materials are regarded as green and are specifically allowed under the provisions of the IRC.
- Green alternatives to conventional materials are available, the selection of which depends on how one defines "green" construction.

CHIMNEYS, FIREPLACES, AND MASONRY HEATERS

This chapter corresponds to IRC Chapter 10, Chimneys and Fireplaces

learning objective

To know and understand how chimneys and fireplaces are generally regulated by the code. Also, to explore green options for fireplaces and masonry heaters and to understand how the IRC requirements impact those choices once made.

IRC CHAPTER 10—OVERVIEW

Chapter 10 of the International Residential Code provides requirements for traditional masonry fireplaces, heaters, and chimneys (see Figure 12-1). It establishes standards for materials used in the construction of fireplaces and provides prescriptive construction methods for fireplaces and chimneys constructed in all climatic regions and seismic zones. Fireplaces, masonry heaters, and chimneys must be safely constructed to ensure that they will exhaust poisonous smoke and gasses safely from the dwelling and that they will not cause a fire due to unsafe construction.

Traditional fireplaces, masonry heaters, and chimneys are typically constructed of solid or hollow masonry units but can also be constructed of concrete or stone (see Figure 12-2). Where hollow masonry is used, it must be grouted solid. In seismic zones D_0, D_1, or D_2, IRC Section R1001.3 requires fireplaces and chimneys to be reinforced with steel rebar to ensure that they do not crumble into pieces in the event of an earthquake. Reinforcing steel is required in both the vertical and horizontal direction and must be interconnected to ensure it will function as intended.

In these same seismic zones, masonry or concrete chimneys must also be anchored to the structure at each floor, ceiling, or roof line that is 6 feet (1,829 mm) or more above grade. This is to ensure that the chimney will remain in place and to prevent it from toppling over should an earthquake occur. Seismic anchorage is not required where the chimney is located completely inside the structure.

Figure 12-1

A masonry heater installed in a dwelling. Note the built-in warming oven above the firebox and the openings near the base for heat distribution.

COURTESY OF RUSS HOLLAND

Seismic anchorage is typically provided by steel straps embedded into the masonry as it is placed. IRC Section R1001.4.1 requires two straps at each connection, embedded at least 12 inches (305 mm) into the chimney and hooked around the outer reinforcing bars. Each strap is then bolted to the floor or roof framing members to ensure a suitable connection with a minimum of two 0.5-inch (13-mm) diameter bolts.

The IRC prescribes methods of construction for all portions of the traditional fireplace including the dimensions of footings and foundations, the dimensions and thicknesses of firebox walls, and criteria for the construction of the throat, lintel, damper, and other components. Because of the shear size and weight of traditional fireplaces, they must be founded on a suitable footing. The IRC requires footings for fireplaces to be 12 inches (305 mm) thick and to extend at least 6 inches (152 mm) beyond the width of the fireplace horizontally. Like all footings, those supporting fireplaces must be founded on native soil suitable for the support of a structure or on engineered fill. Where reinforcing steel is required, it must be tied into the footing as required by Table R1001.1 and Section R609, Grouted Masonry.

Figure 12-2

A traditional wood-burning fireplace. Note the masonry unit construction.

COURTESY OF ISTOCKPHOTO.COM

Construction for fireboxes is also prescribed (see Figure 12-3). For fireplaces lined with 2-inch (51-mm) thick firebrick, masonry forming the walls of the firebox need only be 8 inches (203 mm) thick including the lining. Where fireboxes are unlined, the IRC requires the minimum thickness for the back and side walls to be 10 inches (254 mm) of solid masonry. Fireboxes must be at least 20 inches deep, per Section R1001.6. The throat leading into the passageway above the firebox must be at least 8 inches (203 mm) above the fireplace opening and not less than 4 inches (102 mm) in depth. Additionally, the cross-sectional area of the passageway above the firebox (including the throat, damper, and smoke chamber) must not be less than that of the chimney flue.

Section R1001.5.1 addresses steel fireplace units. These units are typically manufactured from steel and are combined with masonry to form a masonry fireplace. Steel fireboxes must conform to their listing or the requirements of IRC

Figure 12-3

Table R1001.1 Summary of requirements for masonry fireplaces and chimneys.

TABLE R1001.1
SUMMARY OF REQUIREMENTS FOR MASONRY FIREPLACES AND CHIMNEYS

ITEM	LETTER[a]	REQUIREMENTS
Hearth slab thickness	A	4"
Hearth extension (each side of opening)	B	8" fireplace opening < 6 square foot. 12" fireplace opening ≥ 6 square foot.
Hearth extension (front of opening)	C	16" fireplace opening < 6 square foot. 20" fireplace opening ≥ 6 square foot.
Hearth slab reinforcing	D	Reinforced to carry its own weight and all imposed loads.
Thickness of wall of firebox	E	10" solid brick or 8" where a firebrick lining is used. Joints in firebrick ¼" maximum.
Distance from top of opening to throat	F	8"
Smoke chamber wall thickness Unlined walls	G	6" 8"
Chimney Vertical reinforcing[b]	H	Four No. 4 full-length bars for chimney up to 40" wide. Add two No. 4 bars for each additional 40" or fraction of width or each additional flue.
Horizontal reinforcing	J	¼" ties at 18" and two ties at each bend in vertical steel.
Bond beams	K	No specified requirements.
Fireplace lintel	L	Noncombustible material.
Chimney walls with flue lining	M	Solid masonry units or hollow masonry units grouted solid with at least 4 inch nominal thickness.
Distances between adjacent flues	—	See Section R1003.13.
Effective flue area (based on area of fireplace opening)	P	See Section R1003.15.
Clearances: Combustible material Mantel and trim Above roof	R	See Sections R1001.ll and R1003.18. See Section R1001.11, Exception 4. 3' at roofline and 2' at 10'.
Anchorage[b] Strap Number Embedment into chimney Fasten to Bolts	S	$^3/_{16}" \times 1"$ Two 12" hooked around outer bar with 6" extension. 4 joists Two ½" diameter.
Footing Thickness Width	T	12" min. 6" each side of fireplace wall.

For SI: 1 inch = 25.4 mm, 1 foot = 304.8 mm, 1 square foot = 0.0929 m².

Note: This table provides a summary of major requirements for the construction of masonry chimneys and fireplaces. Letter references are to Figure R1001.1, which shows examples of typical construction. This table does not cover all requirements, nor does it cover all aspects of the indicated requirements. For the actual mandatory requirements of the code, see the indicated section of text.

a. The letters refer to Figure R1001.1.

b. Not required in Seismic Design Category A, B or C.

Section R1001. Steel fireplace units sometimes incorporate air chambers for the purpose of allowing air to circulate into the dwelling; some are fan-assisted models. In these cases, the lining must be fully embedded into masonry and circulating air ducts connected to the steel fireplace unit must be constructed of metal or masonry.

The IRC also provides installation criteria for hearths and hearth extensions, clearances to combustible woodwork such as mantles and fireplace surrounds, and detailed requirements for the construction of masonry chimneys, flue linings, and chimney terminations. Because combustible framing is often combined with or located adjacent to masonry fireplaces, appropriate clearances must be provided from the wood framing material to ensure that it does not ignite due to radiant heat. To prevent this from occurring, the IRC requires a 2-inch (51-mm) clearance between the front and sides of the masonry fireplace for all combustible materials and a 4-inch (102-mm) clearance from the back face of the fireplace.

Figure 12-4

The rustic, timeless charm of a fireplace is captured beautifully in this image.

COURTESY OF RUSS HOLLAND

GREEN OPTIONS

From a purely "green" point of view, traditional solid-fuel-burning fireplaces do not make a lot of sense. They consume wood products, a practice that is generally opposed in green thinking. They are remarkably inefficient, in that the vast majority of the heat generated when the wood is burned is lost through the chimney. In addition, smoke and other products of combustion increase air pollution as wood and other fuel types are burned. Regardless, many people enjoy the allure of a wood-burning fireplace with its rustic charm and comfortable, radiant heat. For this reason, the fireplace remains a popular choice in homes (see Figure 12-4).

Is there a green way to incorporate a fireplace into the design of a dwelling? If one carefully plans the type and location of the fireplace, certainly! Despite environmental concerns over traditional wood-burning fireplaces, certain options make some choices greener than others. Air-circulating fireplaces, for example, capture heat that is stored in the firebox and distribute it through ducts, allowing more of the heat generated to be utilized. Rumford fireplaces are generally more efficient than their traditional cousins due to the shallower fireboxes that reflect more heat into the room.

Masonry heaters are similar to air-circulating fireplaces; however, masonry heaters circulate

Figure 12-5

A masonry heater under construction. In this case, clay ducts are installed to facilitate the distribution of heat into the adjacent rooms.

COURTESY OF RUSS HOLLAND

the exhaust gasses generated as the solid fuel burns into heat exchange chambers within the masonry before exhausting it up through the chimney. This process heats the masonry and attached circulating air ducts which convey more heat into the room (see Figure 12-5). Pellet stoves can be thermostatically controlled and are remarkably efficient at space heating and produce considerably less pollution than wood-burning fireplaces, offering another viable option for the green construction project. Fortunately, the IRC allows all of these options as prescriptive choices in the code. Let's explore some of them more fully.

ALTERNATIVE WOOD-BURNING FIREPLACES

As has been the case in other chapters, the IRC provides code criteria for a number of alternative wood-burning fireplaces that can be regarded as green for one reason or another. The following code sections provide criteria for fireplaces that are generally regarded as more efficient or better performing than traditional fireplaces.

Section R1001.5.1 Steel Fireplace Units

The IRC allows steel fireplace units to be incorporated into the masonry fireplace. Such units are convenient and can simplify the construction of the fireplace. Because these types of fireplace units are manufactured, they are required to be tested and listed for their intended use. Thus, in addition to the general code requirements of the IRC, the manufacturer's installation instructions must also be taken into account.

Section R1001.5.1 requires steel fireplace units that incorporate a steel firebox lining to be constructed of steel not less than 0.25 inch (6 mm) thick and be provided with an air-circulating chamber that is vented to the interior of the building. Firebox linings must be encased in solid masonry to provide a total thickness of 8 inches (203 mm). The minimum required thickness of the concrete or masonry itself is 4 inches (102 mm).

Intent

The intent of this IRC section is to allow manufactured, listed steel fireplace units to be used in conjunction with solid masonry to form masonry fireplaces. The code further intends that steel fireplace units be installed in accordance with their listing.

Green Decisions and Limitations

Shall I use a steel, air-circulating type fireplace in my green dwelling? What code requirements will impact this decision?

Steel Fireplace Option

One of the main criticisms of the traditional fireplace is inefficiency. A significant amount of the heat created as the wood burns is lost through the chimney. Air-circulating fireplaces offer a green solution because they capture heat stored in the air chamber and then, through the use of convection or a fan, channel the heat into the room where it can be utilized for space heating (see Figure 12-6). In this way, the heat stored in the thermal mass of the masonry can be captured instead of lost. This makes more efficient use of the solid fuel being burned—a most desirable effect.

Why It's Green

Steel fireplace units with built-in air chambers collect and distribute heat through ducts. Sometimes, this heated air is distributed with the assistance of a fan, which makes for a more efficient use of the fuel burned.

Figure 12-6

An air-circulating, prefabricated steel fireplace unit prior to installation.

COURTESY DELMAR/CENGAGE LEARNING

Section R1001.6 Firebox Dimensions

This IRC section provides minimum firebox depth dimensions. For standard wood-burning fireplaces, again, the minimum required depth is 20 inches (508 mm). Additionally, the IRC requires that the throat be not less than 8 inches (203 mm) above the top of the fireplace opening. According to the 2006 IRC Code and Commentary, Volume I, "the proper functioning of a fireplace is dependent on the size of the face of the opening and the chimney dimensions, which in turn

Why It's Green

Rumford fireplaces have a tall, shallow firebox that allows for more heat to be reflected into the room. More reflected heat means that small fires can provide the same or more heat as large fires in a traditional fireplace. Small fires mean less wood is necessary for a given amount of heat and more efficient performance overall.

are related to the room size." The depth of the fireplace impacts the ability of the firebox to create the draft necessary to pull smoke up and out of the fireplace.

Criteria for the throat location and size are also provided here and are based on the empirical data derived from the successful construction of thousands of fireplaces through the decades. The IRC makes an important exception to the requirements of Section R1001.6 in that it allows for the use of a Rumford fireplace, which has a reduced firebox depth and different firebox dimensions.

Intent

The intent of this section is to ensure that fireboxes for masonry fireplaces are provided with a combustion chamber that will allow for proper draft to safely draw smoke up and out of the fireplace. The code further intends that the fireplace opening dimensions be sized in such a way that the room in which it is provided does not overheat.

Figure 12-7

A Rumford fireplace installed in a dwelling. Note the steeply angled sides and taller, shallower firebox that contribute to more heat distribution into the room.

COURTESY OF BUCKLEY RUMFORD FIREPLACES

Green Decisions and Limitations

Shall I incorporate a Rumford fireplace into the design of my green dwelling?

How does the IRC impact this decision?

The Rumford Fireplace Option

Rumford fireplaces are named after Count Rumford, whose study of the nature of heat and the improvement of the fireplace led to the development of his namesake. Rumford fireplaces were popular from 1796 when he first wrote of them until the mid-1800s. They are characterized by a tall, shallow firebox with steeply angled sides that reflect heat into the room more efficiently than their modern predecessors (see Figure 12-7). The exception to IRC Section 1001.6 allows the firebox depth dimensions to be reduced to 12 inches (305 mm) where the Rumford design is used; however, the firebox opening size and other dimensions must be adjusted accordingly. The Rumford design can be incorporated into the green dwelling to provide a more efficient fireplace system.

TAKE NOTE!

Where the Rumford design is used, the firebox depth can be reduced to 12 inches (305 mm) as discussed; however, the depth must also be at least one-third the width of the fireplace opening, and the throat must be at least 12 inches (305 mm) above the lintel and one-twentieth the cross-sectional area of the fireplace opening. Be sure to meet these requirements when using this exception.

CHAPTER 12 APPLIED TO THE SAMPLE PROJECT

Chapter 12 pertains primarily to masonry fireplaces and chimneys and provides specific criteria for all aspects of their construction. These criteria include the code requirements not only for the footings and supports but also for reinforcement and seismic anchorage. Chapter 12 also addresses steel fireplace units, pellet stoves, and Rumford fireplaces. The following analysis will demonstrate the application of some IRC code provisions for fireplaces with consideration given for the proposed scope of work and our design philosophy, even though the scope of work for the sample project did not specifically include plans for a fireplace or similar unit.

Site Considerations

Let us assume for a moment that we will include a Rumford-style fireplace in the sample project and that it is necessary to locate the fireplace along the 1-hour fire-resistive exterior wall so that we can see how the requirements of the IRC impact our design. Let us also assume that, for simplicity's sake, we will deviate from the use of rammed earth for the construction of the fireplace and utilize conventional building materials for its construction.

Once again, the fire-resistive construction requirements of Section R302.1 will impact the fireplace design due to the placement of the exterior wall of the dwelling 3 feet (914 mm) from the property line. Assuming that the proposed fireplace projects beyond this exterior wall 2 feet (610 mm), a fairly typical occurrence, the outer edge of the fireplace will be located only 1 foot (305 mm) from the property line. Like the exterior wall of the dwelling, the exterior wall of the fireplace parallel to the property line will obviously need to meet the 1-hour fire-resistive construction requirements identified in the code.

A complication occurs, however, from the requirements of Section R1006 of the IRC, which requires a supply of outdoor air to be provided to the firebox to ensure that an adequate supply of oxygen is available for combustion. Recall from earlier chapters in this text that openings of all types (combustion air openings included) are prohibited in walls located less than 3 feet (914 mm) from property lines, thus our design must satisfy three requirements due to site conditions: meet the 1-hour fire-resistive construction requirements of Section R302.1, avoid the placement of any openings in the exterior wall of the fireplace, and provide the required outdoor combustion air per Section R1006.

Building Considerations

The fireplace will utilize conventional, reinforced solid-grouted masonry for the basic construction. Table R1001.1 and Figure R1001.1 provide dimensions for footings, hearths, fireboxes, wall thicknesses, and more, so it will be fairly easy to satisfy the basic code requirements for the design and construction of the fireplace by simply following the prescriptive requirements outlined in the code. The use of reinforced, solid-grouted masonry has added benefits as well, as this will easily satisfy the 1-hour fire-resistive construction requirements of Section R302.1. In fact, reinforced masonry has been used for years in fire-resistive applications and has been demonstrated to meet 2-, 3-, and even 4-hour fire-resistivity tests.

Recall that we opted to use the exception to the requirements of Section R1001.6, Firebox Dimensions, and incorporate the Rumford-style firebox into our design. Thus, we will need to adjust the depth of the firebox to 12 inches (305 mm) and adjust other dimensions accordingly. Again, a fairly simple task because this is specifically allowed by the IRC.

As for the outdoor combustion air requirements outlined in Section R1006, we have a dilemma. We must provide an air passageway from the exterior into the firebox of at least 6 square inches (3,870 mm²) but not greater than 55 square inches (0.035 m²) to ensure that an adequate supply of oxygen will be delivered for combustion. This is an important safety consideration because an inadequate supply of oxygen can cause incomplete combustion as wood burns, generating dangerous levels of carbon monoxide. However, we cannot place the required air intake opening in the exterior wall located parallel to the property line, per Table R302.1. So, what shall we do?

Fortunately, the IRC provides a solution that will enable us to meet all of the requirements of the code. Recall from an earlier chapter that the fire-resistivity requirements for exterior walls and opening restrictions do not apply to walls that are *perpendicular* to the property line. In our case, because the fireplace projects out 2 feet (610 mm) beyond the exterior wall of the dwelling, the short return walls at either side of the fireplace make great candidates for the placement of the required air intake opening because they are perpendicular to the property line. The IRC also allows combustion air to be taken from ventilated crawlspaces; however, this is not an option here because of our concrete slab floor.

CHAPTER SUMMARY

Chapter 10 of the IRC covers chimneys, fireplaces, and heaters constructed in dwellings. It provides standards for materials and methods of installation for masonry chimneys and fireplaces as well as factory-built wood-burning fireplaces. Following are highlights of the chapter.

- The IRC provides criteria for the installation of masonry fireplaces, masonry heaters, and chimneys and allows concrete, solid, and grout-filled hollow masonry or stone to be used.
- Criteria are provided in the text of the code as well as in tabulated form. Clearances to combustibles, hearth dimensions, and other important criteria are identified.
- Masonry fireplaces and chimneys must be reinforced and properly anchored in seismic zones D_0, D_1, and D_2.
- The IRC provides criteria for the installation of masonry heaters.
- The IRC provides criteria for the use of steel fireplace units as well as pellet stoves and other factory-built fireplace units.
- A number of "green" alternatives to the traditional fireplace are outright allowed in the code.
- The manufacturer's installation criteria must be followed as directed by the IRC to ensure a code-compliant installation and to preserve the manufacturer's warranty.

ENERGY CONSERVATION

learning objective

To know and understand how the IRC regulates energy efficiency in dwellings. Also, to explore green options for energy-conserving materials, equipment, and construction methods and to understand how the IRC requirements impact those choices once made.

IRC CHAPTER 11—OVERVIEW

Just as Chapter 3 of the International Residential Code was described as the "meat and potatoes" chapter because of the breadth and depth of the topics covered, so could Chapter 11 of the IRC be described as the "green building" chapter because of its exclusive focus on energy efficiency. Energy conservation is at the heart of green building; and for many, energy conservation *is* green building.

As we have seen, however, green building is about much more than simply saving energy. Green building requires a holistic approach to residential construction that is a product of many steps and considerations. Decisions affecting green building must be made prior to and at every step of a green building project. Still, energy conservation plays a big part in those decision-making processes and must be duly considered.

It is widely accepted that improvements in the energy efficiency of a dwelling—from the performance of the building envelope itself, to the equipment, lighting, and appliances installed within it—contribute to environmental sustainability in a positive way. Thus, a trend has emerged in recent years toward improved energy efficiency in all facets of our lives. For example, dwellings are now routinely constructed to be highly efficient through the use of thick insulation and double-paned windows. Highly efficient compact fluorescent bulbs are now produced in all types and sizes and are widely available. High-efficiency appliances such as clothes washers and dryers are now popular as they offer savings not only by way of reduced energy and water consumption, but also in the form of rebates and tax credits (see Figure 13-1).

Chapter 11 of the IRC addresses energy conservation at all levels, including considerations for the **exterior envelope** (the insulated walls, floors, and ceilings) and

Figure 13-1

An Energy Guide label for an appliance featuring the *Energy Star* logo. In this case, the appliance featured is a refrigerator-freezer.

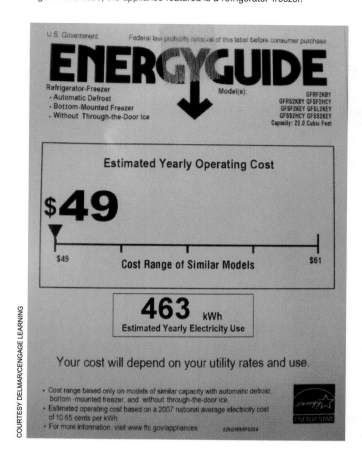

fenestrations (openings in the envelope such as windows and doors), as well as equipment and lighting efficiency. The IRC prescribes minimum insulation levels and performance expectations for all new dwellings, additions, and alterations constructed under its provisions.

BUILDING THERMAL ENVELOPE

In code parlance, the IRC is concerned with the building thermal envelope or, as we will call it, the exterior envelope. The exterior envelope includes the exterior walls and floors, ceilings, and roofs that enclose conditioned space. It also includes the windows and doors, and other openings in the exterior walls. Essentially, the exterior envelope provides a barrier between the heated and/or cooled portions of the dwelling and unconditioned spaces such as attics and crawl spaces or the outdoors. In essence, the building envelope forms a "cocoon" around the conditioned interior spaces of the dwelling with the intention of retaining as much of the conditioned air as possible. The more efficiently the exterior envelope performs this task, the better the quality of the interior space of the dwelling.

Exterior Walls, Floors, and Ceilings

The IRC prescribes levels of performance for building envelopes and the materials used in their construction in Section N1102. Simply put, the IRC prescribes the level of insulation that is needed to achieve a minimum satisfactory level of performance for the exterior envelope. By this point, the familiar theme that the IRC prescribes different requirements based on geographic region should come as no surprise. It stands to reason that the energy and thermal performance demands of a dwelling located in northern Minnesota are very different from those of a dwelling located in southern Florida. Thus, the IRC is equipped to deal with these regional differences through the use of maps, such as the one shown in Figure 13-2.

The IRC divides the United States into climate zones based on a number of factors including average temperature ranges, moisture conditions, and marine influences. Climate zones are presented in a numbering scheme from 1 through 8, with lower numbers generally representing milder climates such as those encountered in the southern and southwestern portions of the United States, and the higher

Figure 13-2

Figure N1101.2
Climate zones, from
the 2009 IRC.

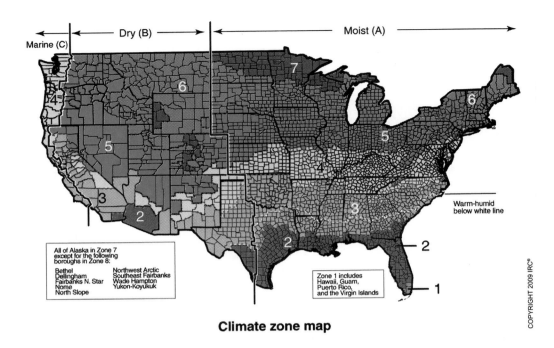

Climate zone map

numbers generally representing colder climates. Of course, it is an oversimplifica-
tion to say that climate exactly follows the numbering scheme presented in the
code. Climates vary considerably even within a particular zone, and we all know
Mother Nature does her best to keep things interesting in terms of weather, no matter where one calls home.

The point here is this: Exterior envelope requirements vary regionally under the
provisions of the IRC. Table N1102.1, in Figure 13-3, shows the insulation and fenestration requirements by component for each of the eight climate zones identified
in the IRC. Insulation values for walls, floors, ceilings, and other similar components
are prescribed in R-values, or "resistance" values. Simply put, R-values are a familiar
way to rate the resistance to the flow of heat for a given level of insulation; the higher
the R-value, the greater the resistance to the flow of heat. Thus, a fiberglass batt with
an R-value rating of 13 would not be as resistant to heat flow as one rated at R-19.

Section N1102.2.4 addresses "mass" walls, which differ from typical frame construction in that the walls are constructed of a solid mass. Mass walls include ICF,
solid concrete and masonry, earth walls, solid timbers, and others. Thus, it is not
possible to install batt or spray-applied foam insulation within the wall because
there is no wall cavity to fill. The IRC allows mass walls and provides criteria for
insulation, which also appears in Table N1102.1 (again, see Figure 13-3).

Fenestrations

Another important consideration for the exterior envelope is the window and door
penetrations. Windows and doors account for, arguably, the single biggest portion of the heat loss in a dwelling; thus, improvements in this area can pay big
dividends when it comes time to pay the utility bill. Like walls and ceilings, the

Figure 13-3

Table N1102.1 Insulation and fenestration requirements by component.

TABLE N1102.1
INSULATION AND FENESTRATION REQUIREMENTS BY COMPONENT[a]

CLIMATE ZONE	FENES-TRATION *U*-FACTOR	SKYLIGHT[b] *U*-FACTOR	GLAZED FENESTRA-TION SHGC	CEILING *R*-VALUE	WOOD FRAME WALL *R*-VALUE	MASS WALL *R*-VALUE[k]	FLOOR *R*-VALUE	BASEMENT[c] WALL *R*-VALUE	SLAB[d] *R*-VALUE AND DEPTH	CRAWL SPACE[c] WALL *R*-VALUE
1	1.2	0.75	0.35[j]	30	13	3/4	13	0	0	0
2	0.65[i]	0.75	0.35[j]	30	13	4/6	13	0	0	0
3	0.50[i]	0.65	0.35[e,j]	30	13	5/8	19	5/13[f]	0	5/13
4 except Marine	0.35	0.60	NR	38	13	5/10	19	10/13	10, 2 ft	10/13
5 and Marine 4	0.35	0.60	NR	38	20 or 13 + 5[h]	13/17	30[f]	10/13	10, 2 ft	10/13
6	0.35	0.60	NR	49	20 or 13 + 5[h]	15/19	30[g]	10/13	10, 4 ft	10/13
7 and 8	0.35	0.60	NR	49	21	19/21	30[g]	10/13	10, 4 ft	10/13

a. *R*-values are minimums. *U*-factors and solar heat gain coefficient (SHGC) are maximums. R-19 batts compressed in to nominal 2 × 6 framing cavity such that the *R*-value is reduced by R-1 or more shall be marked with the compressed batt *R*-value in addition to the full thickness *R*-value.

b. The fenestration *U*-factor column excludes skylights. The SHGC column applies to all glazed fenestration.

c. The first *R*-value applies to continuous insulation, the second to framing cavity insulation; either insulation meets the requirement.

d. R-5 shall be added to the required slab edge *R*-values for heated slabs. Insulation depth shall be the depth of the footing or 2 feet, whichever is less, in zones 1 through 3 for heated slabs.

e. There are no SHGC requirements in the Marine Zone.

f. Basement wall insulation is not required in warm-humid locations as defined by Figure N1101.2 and Table N1101.2.

g. Or insulation sufficient to fill the framing cavity, R-19 minimum.

h. "13+5" means R-13 cavity insulation plus R-5 insulated sheathing. If structural sheathing covers 25% or less of the exterior, R-5 sheathing is not required where structural sheathing is used. If structural sheathing covers more than 25% of exterior, structural sheathing shall be supplemented with insulated sheathing of at least R-2.

i. For impact-rated fenestration complying with Section R301.2.1.2, the maximum *U*-factor shall be 0.75 in zone 2 and 0.65 in zone 3.

j. For impact-resistant fenestration complying with Section R301.2.1.2 of the *International Residential Code,* the maximum SHGC shall be 0.40.

k. The second *R*-value applies when more than half the insulation is on the interior.

COPYRIGHT 2009 IRC®

IRC prescribes requirements for fenestrations to ensure that windows and doors perform to a certain minimum level. Unlike walls, windows and doors are rated with U-factors (see Figure 13-4).

U-factors are a measure of the *conductivity,* or the rate of heat transfer, for a particular surface or assembly as opposed to resistance to the flow of heat. U-factors are the reciprocal of R-values and are typically presented as a decimal number. To appreciate how much less insulative value a window has than an **opaque wall** (a solid portion of a wall containing no windows or doors), consider the following:

The table in Figure 13-3 shows that a U-factor of 0.35 is required for fenestrations in Climate Zone 4 (except marine). For the sake of comparison, let's convert this number to an R-value. To convert this number to an R-value, take its reciprocal; that is, take the number 1 and divide it by 0.35. The resulting number, 2.86, is the comparative R-value for the window. As we can see, even windows that perform well (U-factor of 0.35) are considerably less resistant to heat flow than an opaque wall (R-21 for the wall versus R-2.86 for the window), accounting for significantly more heat loss.

Figure 13-4

An example of a typical manufacturer's label from a window installed in a dwelling. Note the window performance ratings on the right side of the label.

Air leakage at fenestrations is also regulated under the provisions of the code. Section N1102.4.4 establishes a maximum air infiltration rate for windows, skylights, and sliding glass doors of 0.3 cubic feet per minute per square foot (1.5 (L/s)/m^2) of glazed area. Swinging doors are also provided with a maximum air infiltration rate of 0.5 cubic feet per minute per square foot (2.5 (L/s)/m^2) of door area. In so doing, the IRC further attempts to regulate the performance of windows and doors by controlling the amount of unconditioned air that can enter the dwelling through the fenestration.

Building Systems

Building systems are also regulated under the provisions of Chapter 11. IRC Section N1103 provides criteria to ensure that heating and cooling systems are minimally energy efficient. Where heat pumps are utilized and also provided with electric-resistance heating as a supplementary heat source, controls must be provided that prevent supplemental heat operation when the heat pump can meet the heating demand, per Section N1103.1.2.

As heated or cooled air travels through a duct, it warms or cools, respectively. Ducts are often routed through unconditioned spaces (e.g., ventilated attics and crawl spaces), which are subject to extremes of temperature and thus accelerate heat loss or gain. Add to this the fact that at least some leakage occurs at all duct joints, and it will come as no surprise that a large amount of heat is lost through duct systems. The IRC establishes energy performance criteria in Section N1103.2 in an attempt to reduce this heat loss.

Ducts that supply conditioned air throughout the dwelling and are routed through unconditioned spaces must be insulated to R-8 to minimize heat loss as the conditioned air travels through them to the outlet (see Figure 13-5). Other

Figure 13-5

An insulated duct installed in an unconditioned attic space. Note the R-8 insulation label.

COURTESY DELMAR/CENGAGE LEARNING

ducts, such as those used to return air, must be insulated to R-6. Additionally, ducts, air handlers, filter boxes, and other HVAC components must be sealed to prevent conditioned air from leaking through openings per Section M1601.4.

Additionally, the piping in mechanical systems and in circulating hot water systems must be insulated. Again, this is to reduce the amount of heat lost through exposed piping. The IRC specifies a minimum R-value of R-3 for mechanical system piping insulation capable of carrying fluids at temperatures below 55° F (13° C) and greater than 105° F (40° C). R-2 insulation is required for all piping in circulating hot water systems. The 2009 IRC also established new lighting efficiency requirements for permanently installed lighting fixtures.

GREEN OPTIONS

Considering that this chapter of the IRC is all about green construction, is it possible to get any "greener"? Yes, of course, when you consider the simple measures and steps that can be taken in the planning of a dwelling to improve energy efficiency.

Some of these methods have already been explored, such as the unvented conditioned attic assemblies discussed in Chapter 10 of this text. Other measures are worth further exploration here. For example, simple improvements in the performance of the exterior envelope can reduce heat loss and reduce infiltration. Window and door performance can be enhanced through the careful selection of windows and doors that are best suited for use in a particular climate zone. Building systems performance can also be improved by routing ducts through conditioned areas, carefully sealing all connections, and selecting high-efficiency HVAC system components. The following sections explore each of these areas more thoroughly.

ENERGY CONSERVING MEASURES

The IRC attempts to conserve energy holistically, that is, by considering "the big picture." It regulates energy consumption across a broad spectrum of categories, as discussed in the following sections, rather than in one category or another. In so doing, the code utilizes a "systems" approach. The dwelling can be thought of as a system of interrelated parts, each dependent on the other for energy efficiency.

Section N1102.4.1 Building Thermal Envelope

This IRC section requires the building envelope to be sealed to limit infiltration, which, if ignored, can significantly increase energy consumption. Air infiltration can cause a room to feel drafty and uncomfortable, which in turn prompts the thermostat to be inched ever upward to compensate, consuming ever more energy. Air infiltration can occur at just about every seam, joint, and intersection within a dwelling, so it is important to thoroughly seal all areas as prescribed by the code.

The IRC requires sealants at all joints, seams, and penetrations, at openings between window and door assemblies and their jambs, at utility penetrations, and at a host of other areas. New in the 2009 IRC is the addition of the attic access hatch and the rim joist junction to the list, both of which have been identified as sources of potentially significant amounts of air infiltration. The IRC allows caulk, gaskets, weatherstripping, and other approved sealants to be used.

Why It's Green

Taking extra measures to seal the exterior envelope can significantly reduce air infiltration. In extreme climates, this can potentially reduce energy consumption, because less energy is needed to counteract the effects of the infiltration. Using less energy for heating and cooling is generally regarded as a sustainable practice.

Intent

The intent of this IRC section is to increase energy efficiency by reducing the amount of air infiltration that can occur in the dwelling. The code further intends that sealants and sealant systems used for this purpose be suitable for the task and be of approved materials.

Green Decisions and Limitations

Are there other sources of air infiltration that I should consider sealing?

Envelope Sealing Option

Although the list provided in the IRC is extensive, it does not identify all possible areas where infiltration could occur. Air infiltration can occur at electrical outlets and switches, recessed lighting fixtures, wall sill plates, around fireplaces (especially prefabricated and direct vent factory-built models), at cable penetrations through the floor, and more (see Figure 13-6). In short, air infiltration can occur just about anywhere there is a penetration or hole into or through a wall, floor, roof, or ceiling component.

TAKE NOTE!

Where sealants, gaskets, or other air-barrier systems will be placed in contact with heat-generating sources (e.g., recessed light fixtures), be sure that they are rated and approved for such an application. The quest to save energy by controlling infiltration should never be placed ahead of fire safety.

Figure 13-6

A diagram showing typical sources of air infiltration. Sealing these infiltration areas provides increased energy efficiency.

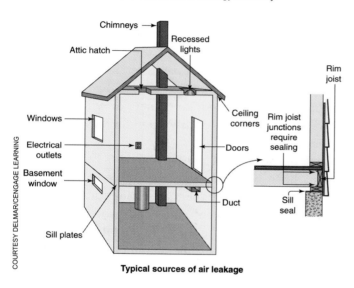

COURTESY DELMAR/CENGAGE LEARNING

Typical sources of air leakage

Section N1102.3 Fenestrations

Windows and doors perform rather poorly when compared to the opaque walls in which they are installed. As we saw in the earlier example, even windows that perform relatively well are still very low in comparison to walls. Thus, windows and doors represent an area where even modest gains in performance can make a big difference in energy efficiency.

The IRC attempts to regulate fenestrations in a variety of ways. As we saw earlier, fenestrations are required to meet certain minimum levels of performance. In milder climates (Climate Zones 1 through 3), a U-factor of 1.2 to 0.5 is acceptable (that's R-0.83 to R-2). In colder climates (Zones 4 through 8), a U-factor of 0.35 is required (R-2.86). The U-factor, however, is not the only consideration affecting fenestrations.

In hot climates such as in the South and Southwest, where cooling is often a bigger concern than heating, the effects of solar gain are unwelcome. Significant heating load is introduced to a dwelling through passive solar gain which in turn takes energy to cool to maintain comfortable indoor temperatures. Greater heating load within a dwelling means more energy is necessary to cool it.

The IRC attempts to regulate solar heat gain within a dwelling by establishing a requirement for glazed fenestrations (windows) to meet certain Solar Heat Gain Coefficient (SHGC) criteria (refer to Figure 13-3). SHGC is a measure of how much of the sun's solar energy is admitted through the window. The SHGC of 0.35 established in the table means that 35% of the sun's solar energy is admitted through the window. Windows with a lower SGHC admit less solar energy than those with a higher rating. This is accomplished by special reflective coatings placed on the panes of glass when the window is manufactured to minimize solar gain. Note that the SHGC is "not required" in colder climates where solar heat gain is actually desirable.

Why It's Green

Windows and doors with U-factors lower than the code-required minimums perform more efficiently because they convect less heat (remember, heat flows from hot to cold). Because windows and doors represent a source of great heat loss, if not *the* greatest source, gains in this area can potentially reduce energy consumption significantly. In areas where solar gain is undesirable, windows with a lower SHGC can improve thermal performance by reducing the amount of solar energy that enters through the glazing.

Intent

The intent of this IRC section is to establish minimum levels of performance for energy efficiency in windows and doors. The code further intends that fenestration performances vary by geographic location. The IRC also intends that glazed fenestrations meet SHGC requirements in Climate Zones 1 through 3.

Green Decisions and Limitations

Shall I incorporate windows and doors with lower U-factors and/or greater SHGC in the design of my green dwelling?

Improved U-Factor Option

Windows and doors with lower U-factors make sense when they are readily available. Obviously, windows and doors that perform better may cost more, so a cost-to-benefit analysis must be performed to ensure that they make sense financially. The rate of return from energy savings versus the additional costs for the windows should be examined closely. In areas where temperatures routinely reach extremes or where energy costs are particularly high, they might make better sense than, say, in a moderate climate or where energy costs are comparatively low. The local utility provider may be able to assist in this analysis.

Still, windows that perform in the U-0.31 or U-0.32 range are readily available and have become quite affordable. Technology has improved significantly in recent years as gasses such as argon and krypton have been incorporated into the window design. Vinyl and vinyl-clad windows perform better than metal frame windows because vinyl is not as good a conductor of heat as metal. So, it can be relatively easy to improve window performance simply by selecting windows or doors with lower U-factors, and may cost the same as windows of average performance.

TAKE NOTE!

U-factors for windows, doors, and skylights must be determined in accordance with the National Fenestration Rating Council (NFRC) Standard 100 by an accredited, independent laboratory and must be labeled and certified by the window manufacturer. SHGC for glazed windows, doors, and skylights must be determined in accordance with NFRC Standard 200 by an accredited, independent laboratory and must be labeled and certified by the window manufacturer. Be certain that all windows, doors, and skylights considered for installation in your green dwelling meet these standards.

Section N1103.7 Snow Melting System Controls

Having shoveled countless feet of snow in my youth, there is perhaps no more brilliant idea than to heat the sidewalks and driveways around one's home to control snow. Even though one's chiropractor might also think a snow melting system is a brilliant idea for maintaining a healthy back, from the "energy conservation" point of view, not so much. After all, these systems heat the *outside*! More than a few of you reading this can recall a moment from childhood, when your parents hollered, "Are you going to heat the whole outside?" after you left a door open in winter. Consequently, you may be raising your eyebrows at this moment.

Snow melting systems actually make sense in areas where enormous amounts of snow fall each year, such as in mountainous regions in the western states or where there is a desire to eliminate chemical snow melt and deicing agents. Snow melt systems also make sense for persons with physical limitations that prevent physical snow removal (i.e., shoveling) or where it is necessary to maintain a clear path at all times for safety reasons. Thus, these types of systems are becoming more common.

Snow melt systems use approved hot water tubing embedded in concrete sidewalks and driveways to raise the temperature of the concrete surface to greater than 32° F (0° C). In the 2009 IRC, a new addition requires controls for snow melt systems in an effort to limit energy consumption (see Figure 13-7). For those systems provided with heated water from an energy service inside the building, automatic controls capable of shutting off the system must be provided when the pavement temperatures exceed 50° F (10° C) or where no precipitation is falling. An automatic or manual control must also be provided to shut off the system when outdoor temperatures rise above 40° F (5° C).

Intent

The intent of this IRC section is to limit the amount of energy used to operate snow melt systems that are supplied through energy service to the building. The code further intends that these systems be provided with automatic controls to ensure that they will not continue to consume energy when there is no longer a need for them to operate. The code also intends that these systems be provided with a manual or automatic shut-off switch.

Green Decisions and Limitations

Shall I incorporate other system controls similar to that required for snow removal systems?

What code implications exist that will impact this decision?

Figure 13-7

A diagram showing the layout of a snow melt system and controls.

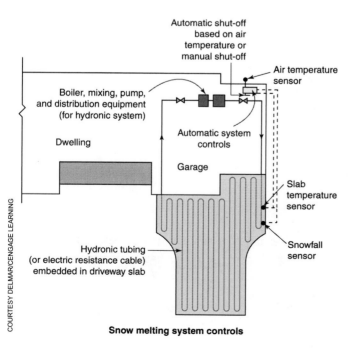

COURTESY DELMAR/CENGAGE LEARNING

Automatic shut-off based on air temperature or manual shut-off

Air temperature sensor

Boiler, mixing, pump, and distribution equipment (for hydronic system)

Automatic system controls

Dwelling

Garage

Slab temperature sensor

Snowfall sensor

Hydronic tubing (or electric resistance cable) embedded in driveway slab

Snow melting system controls

Why It's Green

System controls such as the one described here help to limit energy use because they can be set up to operate only when needed, rather than around the clock. Improvements in and the availability of technology make many of these controls "smart," in that they can recognize when the systems are needed without human intervention. Systems that operate automatically and only when necessary conserve energy.

Options

Where increased energy savings is desired through the use of system controls, consider the following green options.

Other system controls: Automatic controls can be provided for a number of different systems within the house. The first and most common is the thermostat used to control the main HVAC system (see Figure 13-8). Programmable thermostats can be obtained for a nominal cost at most home improvement centers; they can easily be programmed to maintain higher indoor temperatures when the dwelling is occupied and lower temperatures when empty. Some controls might be available through the local utility company to control traditional tank-type water heaters during peak times rather than 24 hours per day. Other automatic controls that should be considered are irrigation timers, swimming pool heaters, and occupancy sensors for lights.

High-efficiency lighting: Beginning with the 2009 edition of the IRC, at least 50% of the bulbs installed in permanent lighting fixtures for new construction are required to be compact fluorescent bulbs or other "high-efficacy" **lamps** (bulbs). Other bulbs that meet the definition of high efficacy are T-8 or smaller tube fluorescent bulbs or lamps with the following ratings.

- 60 lumens per watt for bulbs over 40 watts
- 50 lumens per watt for bulbs over 15 watts to 40 watts
- 40 lumens per watt for bulbs 15 watts or less

Figure 13-8

A programmable thermostat installed in a dwelling. Note the programming instructions provided on the inside cover.

COURTESY DELMAR/CENGAGE LEARNING

Figure 13-9

A stylish and contemporary bathroom sink that utilizes blue LED lighting for accent. LED lighting is available in white and a variety of other colors.

COURTESY OF ISTOCKPHOTO.COM

Consider replacing all possible bulbs with high-efficacy lamps to see real energy savings. For example, a standard 75 watt bulb can be replaced by a 13 watt compact fluorescent bulb with equivalent brightness. Multiply this times 40 or 50 bulbs per dwelling and the overall energy savings can be significant. Compact fluorescent lamps are not the only means by which this code requirement can be met. Other high-efficacy lamp types include light emitting diodes (LEDs) and high-efficiency incandescent lamps (see Figure 13-9). Be sure to verify the lumens-per-watt ratio for lamps used in new construction.

CHAPTER 13 APPLIED TO THE SAMPLE PROJECT

Chapter 13 pertains to energy conservation in one- and two-family dwellings. It provides minimum insulation levels and performance criteria for building envelopes, fenestrations, building systems, and lighting. Additionally, Chapter 13 addresses air infiltration and other topics intended to improve the thermal performance of buildings. The following analysis will demonstrate the application of some IRC code provisions for energy conservation with consideration given for the proposed scope of work and our design philosophy.

Site Considerations

The first issue we must address as it relates to energy conservation is the location of the site in a given climate zone. Up to this point, we have been speaking generically about the sample project so that we could apply the IRC code provisions broadly. However, because the energy conservation requirements of the IRC are so specific and dependent upon the specific climate zone in which the project is located, we must now assign a climate zone to apply the code. For this exercise, we will assume the climate zone is in the middle of the range, Climate Zone 4 (not Marine 4). See Table N1102.1 in Figure 13-3 for specific requirements in this climate zone.

Building Considerations

Table N1102.1 identifies insulation and fenestration requirements by component for all aspects of the sample project. Fortunately, the IRC provides criteria for mass walls so it is easy to determine the required performance of the rammed-earth walls we have specified for our project; in this case, the required R-value is R-10, since more than half of the insulation will be provided on the inside of the dwelling (see footnote k). Thus, our AMM will need to demonstrate that the proposed rammed-earth walls will meet at least the R-10 requirement. Given the massive nature of rammed-earth walls, it is likely we will far exceed this, which is a good thing considering we want our dwelling to be as energy efficient as possible.

Our ceilings must be insulated to R-38 per the table; but these values apply to wood framing. Recall that we will be framing our roof-ceiling with light-gauge steel framing, so we may have to adjust the insulation levels to compensate for additional heat loss. As we suspected, this is in fact the case per IRC Section N1102.2.5. Fortunately, the IRC provides a user-friendly table (not shown) that tells us we must adjust the insulation from R-38 to R-49 to account for the use of steel framing. Also recall from Chapter 10 of this text that we opted to use an unventilated attic with air-impermeable insulation, so the required R-49 insulation levels must be achieved through the use of air-impermeable insulation that is sprayed to the underside of the roof sheathing. Foam plastics are heavily regulated under the provisions of the code, so we will want to take care in selecting a product that will meet not only the energy-efficiency requirements of IRC Chapter 11 but also the strict requirements of IRC Sections R302.10 and R316 *at the required thickness.*

A quick review of this table shows that fenestrations must have a maximum U-factor of 0.35. Recall that, in our design philosophy, we said we want the project to be as energy efficient as reasonably possible. The scope of work also said that we would select doors and windows for maximum energy efficiency, so we will want to select windows that perform better than the required U-0.35 (this means the U-factor for our windows must be *less* than 0.35). Fortunately, many fine, high-efficiency window and door choices are readily available and we settle on windows with a U-factor of 0.31, which will easily meet the code requirements. No SHGC is required in this climate zone.

While it might appear that Table N1102.1 requires slab insulation of R-10 to a depth of 2 feet (610 mm), IRC Section N1102.2.8, Slab-On-Grade Floors, tells us that this requirement applies only where the floor surface is located 12 inches (305 mm) below grade. The remainder of the table is not applicable to this project.

CHAPTER SUMMARY

Chapter 11 of the IRC addresses the various code requirements for energy conservation in dwellings. It provides standards for materials and requirements for construction methods to ensure energy efficiency in dwellings. The entire chapter is devoted to an inherently sustainable topic—energy efficiency. Following are highlights of the chapter.

- Chapter 11 of the IRC provided energy conservation criteria for the building's thermal envelope, fenestrations, and building systems.
- The sole focus of IRC Chapter 11 is energy conservation.
- Chapter 11 of the IRC provides prescriptive code requirements for wall, floor, roof, and ceiling insulation for both wood and steel frame structures.
- The IRC prescribes minimum thermal performance criteria for fenestrations (windows, doors, and skylights).
- The IRC prescribes SHGC criteria for glazed windows.
- U-factors and other performance criteria for fenestrations must be determined per NFRC Standards 100 and 200, through tests performed by independent, accredited laboratories.
- The IRC establishes requirements for controls for certain energy-consuming systems installed in dwellings.
- High-efficacy lighting is required in 50% of the permanent lighting fixtures installed in new construction.
- The manufacturer's installation guidelines for components must be observed.

MECHANICAL EQUIPMENT

14

This chapter corresponds to Part V of the IRC, Mechanical, which includes selected material from IRC Chapters 12 through 23. This chapter also corresponds to Part VI, Fuel Gas, which includes selected material from IRC Chapter 24.

learning objective

To know and understand how mechanical equipment and systems are generally regulated by the code. Also, to explore green options for mechanical equipment, HVAC, and related systems and to understand how the IRC code requirements impact those choices once made.

IRC CHAPTERS 12 THROUGH 24—OVERVIEW

Unlike other chapters in this text, Chapter 14 addresses a range of International Residential Code chapters generally devoted to the topic of mechanical systems. Broadly, this chapter will address topics including general mechanical code requirements, heating and cooling equipment, ducts, exhaust systems, special equipment, hydronic heating systems, and more. In short, it will address all things mechanical as they relate to the IRC.

In this chapter, we will use the acronym HVAC (heating, ventilation, and air conditioning) to reference mechanical systems in general; and when it appears, it will generally include all components of the mechanical from the heating and cooling equipment to the ducts and everything in between. Where specific HVAC equipment is addressed, it will be referred to as an appliance and the text will make it clear what component is being referenced for clarity.

Chapters 12 through 24 of the IRC are devoted to the topic of mechanical systems that are permanently installed within a dwelling. Therefore, portable mechanical equipment such as window-mounted, plug-in type air conditioning units and similar items would not be covered under the scope of the IRC.

Mechanical systems are used to condition (heat or cool) the air inside the dwelling so that it is safe and comfortable (see Figure 14-1). As we discussed earlier, in cold climates (those with a winter design temperature below 60° F, or 16° C), a dwelling must be provided with heating facilities to maintain the code-required indoor temperature of 68° F (20° C) at the required outdoor design temperature. It is important to point out that the IRC does not require a structure to be *cooled* to

Figure 14-1

A modern gas furnace installed in the garage of a townhouse. Note the required gypsum wallboard placed on the walls *prior* to the furnace installation to ensure a full separation between the garage and the living areas.

COURTESY DELMAR/CENGAGE LEARNING

maintain the required 68° F (20° C) temperature outlined in Section R303.8, Required Heating. Thus, in warm climates, these code provisions cannot and should not be interpreted to mean that cooling is required for any dwelling.

A BRIEF LESSON IN THERMODYNAMICS

To fully understand the way mechanical systems operate, it is first necessary to understand some of the basics of **thermodynamics**, or the study of heat and energy. One of the simplest but most important concepts to understand is that heat flows from hot to cold. In other words, heat will flow from a warm environment toward a cooler environment until the temperatures reach equilibrium. This concept is called the *zeroth law of thermodynamics,* which states that when two systems come into contact with each other, there will be an exchange of energy until the systems reach thermal equilibrium.

To prove this point, imagine for a moment that you pour very hot coffee into a ceramic mug. You set the mug on the kitchen counter to let it cool for a moment but then you get distracted and come back 1 hour later. Will the coffee be *hotter* than when you left it? Of course not! In fact, it will not only be much colder than it was originally, but it will likely be the same or very close to the air temperature in the room and it will stay at that room temperature until the air temperature in the room changes. We call this stabilized temperature **thermal equilibrium.** The heat stored in the coffee simply flowed through the walls of the mug and into the room (exchanged from hot to cold) until the temperature of the coffee reached equilibrium with the room temperature (see Figure 14-2).

Figure 14-2

A sketch showing heat flow from a warm environment to a cold environment.

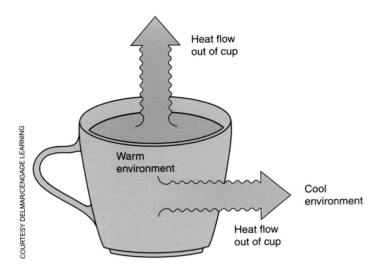

Heat flow out of cup

Warm environment

Cool environment

Heat flow out of cup

Now imagine the same experiment but this time the coffee is placed in an insulated, closed travel mug instead of the ceramic mug. The same hour elapses, but what do you notice about the temperature of the coffee in the travel mug when you return? Of course, the temperature is much closer to when you left it. The coffee may have cooled a bit, but it is likely to be much warmer than the coffee in the ceramic mug, because of the insulating properties and the closed system involved with the travel mug.

The insulated walls of the travel mug greatly slow the exchange of heat; that is, the insulation in the walls of the travel mug resists the flow of heat. The closed system also contributes to the retention of heat because it traps the steam that would otherwise have escaped into the room. Recall from the last chapter that this "resistance to the flow of heat" is the same effect that insulation has in the walls of a dwelling. The *rate* at which the heat flows through the walls of the insulated mug is much slower than that of the uninsulated mug. This *rate of heat transfer* correlates with the U-factor, as discussed in the last chapter for fenestrations.

Now, let us expand this thinking to a dwelling in a cold climate. Like the coffee mug examples, the temperature inside the dwelling represents the warm system and the outdoor temperature represents the cold system. The insulated walls, floors, and ceilings that enclose the conditioned space are essentially the equivalent of the insulated travel mug. The more tightly closed and the better insulated the exterior envelope is, the longer the dwelling will retain the warm temperatures inside. It makes sense, then, that the better we seal the thermal envelope to prevent infiltration, the more insulation we provide.

Still, no system is perfect, so no matter how well we insulate the walls or seal the envelope, heat will still flow out of the dwelling and into the cold outdoors. This is where the mechanical system comes into play (see Figure 14-3). The mechanical system (i.e., the furnace, heat pump, or similar) replaces the heat that is lost through the envelope. That is, the mechanical system adds *energy* to the system in the form of heat, by burning a fuel such as fuel oil, propane, or natural gas or through the use of electricity. The heat that is generated from the furnace or other heating device is distributed throughout the dwelling via ducts or other means.

Throughout the winter months, this cycle repeats again and again. The building owner sets the thermostat to maintain a certain temperature. This action calls for heat to be added to the dwelling, triggering the furnace to come on. Once the desired temperature is reached, the furnace shuts off. Heat flows from the warm space through the building envelope to the cold outdoors, lowering the indoor temperature. This triggers the thermostat to activate, calling for heat from the furnace which starts the cycle all over again.

Figure 14-3

A sketch showing heat flow out of a dwelling during winter months when temperatures are colder outside than inside.

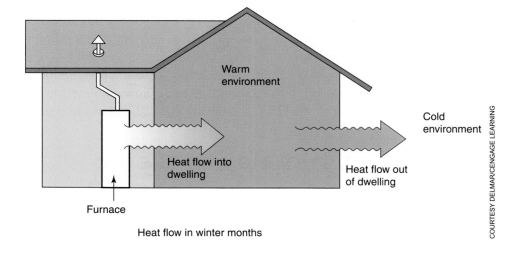

Heat flow in winter months

Figure 14-4

A sketch showing heat flow into a dwelling during summer months when temperatures are warmer outside than inside.

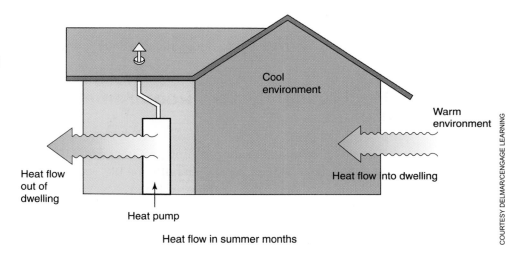

Heat flow in summer months

The opposite is true in warm climates. In keeping with the convention that heat flows from hot to cold, warm outdoor temperatures cause heat to flow through the building envelope *into* the cooler space inside the dwelling. This action progressively warms the interior space. In buildings without some sort of cooling system, indoor temperatures rise until they generally reach equilibrium with the outdoors. Buildings provided with some sort of cooling system, such as an air conditioner, evaporative cooler, or heat pump, cool the space, effectively removing heat from the interior space (see Figure 14-4). This lower temperature inside the dwelling causes heat to flow into the space and again, as we saw in the heating example, the cycle repeats again and again throughout the warm months.

Mechanical systems must be properly sized to ensure that they can quickly and efficiently replace the heat lost in the space. These systems must remove moisture from the air as well. Systems that are improperly sized can cause problems in a variety of ways. If they are too large, they might run in short, rapid cycles which might not run long enough to properly dry the air and consume a great deal of energy when starting and stopping. Systems that are too small might run constantly and never produce enough heat to adequately condition the space. Either way, these inefficiencies increase operating costs (sometimes drastically) or can cause excessive wear on the mechanical equipment, potentially increasing maintenance costs.

With this background, we are now prepared to examine the various ways in which the IRC regulates mechanical systems and their components.

MECHANICAL SYSTEMS AND THE IRC

Unlike previous chapters in this text, Chapter 14 covers a range of chapters in the IRC that collectively make up the code requirements for mechanical systems. For clarity and to ensure a comprehensive treatment of this topic in the space allotted, only pertinent chapters are explored. The material that is necessary for consideration in the green dwelling is presented here; however, the code sections presented in this chapter are by no means a complete representation of all of the code requirements. Consult the 2009 edition of the International Residential Code for a complete treatment of the applicable code provisions.

General Mechanical System Requirements: IRC Chapter 13

The IRC provides the same scoping provisions for HVAC equipment and systems as it does in other chapters. Generally, the IRC requires HVAC systems to comply with one or more of the chapters in Part V, Mechanical, and/or Part VI, Fuel Gas. Where systems are not specifically addressed in the IRC, they are required to meet the requirements of the 2009 International Mechanical Code and the 2009 International Fuel Gas Code (where adopted).

All HVAC equipment is required to be listed and labeled per the provisions of Section M1302. Further, it must be listed for the application in which it is installed and used, unless it has been approved by the code official under the provisions of IRC Section R104.11, Alternative Materials, Design and Methods of Construction and Equipment. This basically means that a particular piece of equipment must be specifically approved for the use intended. For example, if a fireplace stove regulated under the provisions of Section M1414 is to be installed in an alcove, the model chosen must be specifically listed for an alcove installation (see Figure 14-5).

IRC Section M1305 requires mechanical appliances to be provided with access for inspection, repair, and maintenance. The code requires ready access without the necessity to remove permanent construction (a wall, for example) or other appliances to service the equipment. Required clearances vary depending on the particular type of equipment and their location; however, generally a level working space 30 inches (762 mm) by 30 inches (762 mm) must be provided in front of the control side to service the appliance.

Among other things, the IRC addresses access and safety requirements for appliances installed in a variety of locations such as in rooms, in attics, and under floors.

Figure 14-5

A wood stove installed in an alcove. Wood stoves must be specifically listed for installation in an alcove.

COURTESY OF MOUNTAIN LOG HOMES OF COLORADO, INC.

Where appliances are installed in attics or under floors, the IRC requires a clear passageway within and an access opening into the space large enough to remove the largest appliance. Additionally, the IRC requires appliances located within garages and having an ignition source to be elevated such that the source of ignition (flame or spark) is 18 inches (457 mm) above the garage floor. This important safety requirement is intended to prevent gasoline vapors that accumulate at the floor from being ignited.

This chapter also requires mechanical appliances to have proper clearances from combustible materials (those that ignite and burn when exposed to heat or flame), to ensure that heat-generating appliances will not ignite construction materials. These clearances are established by the appliance manufacturer in the listing. Required clearances may be reduced where protection of the combustible material is provided in accordance with the appliance manufacturer's instructions and Table M1306.2. Allowed reductions are based on the form of protection provided (see Figure 14-6).

Heating and Cooling Equipment: IRC Chapter 14

This chapter of the IRC provides specific code criteria for virtually all forms of heating and cooling equipment. Heat pumps, central furnaces, radiant heating systems, room heaters, and others are addressed within this IRC chapter. Clearances must be maintained to all equipment so that blowers, filters, motors, and other components can be serviced.

Section M1401.3 requires heating and cooling equipment to be properly sized to serve the entire building in which it is installed, per the Air Conditioning Contractors of America (ACCA) Manual S, based on building loads calculated in accordance with ACCA Manual J. These publications have become the industry standards for sizing HVAC systems in residential construction. As discussed previously, it is imperative that HVAC systems be properly sized to work efficiently and to perform their required duty.

IRC Chapter 14 provides specific, relevant installation criteria for each type of appliance. For example, the IRC requires vented wall furnaces to be located where they will not cause a fire hazard to doors, furnishings, and other combustible materials in M1409.2. This section also requires vented wall furnaces to be located so that doors will not swing within 12 inches (305 mm) of the air inlet or outlet.

Figure 14-6

Table M1306.2 Reduction of clearances with specified form of protection.

TABLE M1306.2
REDUCTION OF CLEARANCES WITH SPECIFIED FORMS OF PROTECTION[a, c, d, e, f, g, h, i, j, k, l]

TYPE OF PROTECTION APPLIED TO AND COVERING ALL SURFACES OF COMBUSTIBLE MATERIAL WITHIN THE DISTANCE SPECIFIED AS THE REQUIRED CLEARANCE WITH NO PROTECTION (See Figures M1306.1 and M1306.2)	WHERE THE REQUIRED CLEARANCE WITH NO PROTECTION FROM APPLIANCE, VENT CONNECTOR, OR SINGLE WALL METAL PIPE IS:									
	36 inches		18 inches		12 inches		9 inches		6 inches	
	Allowable clearances with specified protection (Inches)[b]									
	Use column 1 for clearances above an appliance or horizontal connector. Use column 2 for clearances from an appliance, vertical connector and single-wall metal pipe.									
	Above column 1	Sides and rear column 2	Above column 1	Sides and rear column 2	Above column 1	Sides and rear column 2	Above column 1	Sides and rear column 2	Above column 1	Sides and rear column 2
3 ½-inch thick masonry wall without ventilated air space	—	24	—	12	—	9	—	6	—	5
½-in. insulation board over 1-inch glass fiber or mineral wool batts	24	18	12	9	9	6	6	5	4	3
Galvanized sheet steel having a minimum thickness of 0.0236-inch (No. 24 gage) over 1-inch glass fiber or mineral wool batts reinforced with wire or rear face with a ventilated air space	18	12	9	6	6	4	5	3	3	3
3½-inch thick masonry wall with ventilated air space	—	12	—	6	—	6	—	6	—	6
Galvanized sheet steel having a minimum thickness of 0.0236-inch (No. 24 gage) with a ventilated air space 1-inch off the combustible assembly	18	12	9	6	6	4	5	3	3	2
½-inch thick insulation board with ventilated air space	18	12	9	6	6	4	5	3	3	3
Galvanized sheet steel having a minimum thickness of 0.0236-inch (No. 24 gage) with ventilated air space over 24 gage sheet steel with a ventilated space	18	12	9	6	6	4	5	3	3	3
1-inch glass fiber or mineral wool batts sandwiched between two sheets of galvanized sheet steel having a minimum thickness of 0.0236-inch (No. 24 gage) with a ventilated air space	18	12	9	6	6	4	5	3	3	3

For SI: 1 inch = 25.4 mm, 1 pound per cubic foot = 16.019 kg/m³, °C = [(°F) − 32/1.8], 1 Btu/(h × ft² × °F/in.) = 0.001442299 (W/cm² × °C/cm).

a. Reduction of clearances from combustible materials shall not interfere with combustion air, draft hood clearance and relief, and accessibility of servicing.

b. Clearances shall be measured from the surface of the heat producing appliance or equipment to the outer surface of the combustible material or combustible assembly.

c. Spacers and ties shall be of noncombustible material. No spacer or tie shall be used directly opposite appliance or connector.

d. Where all clearance reduction systems use a ventilated air space, adequate provision for air circulation shall be provided as described. (See Figures M1306.1 and M1306.2.)

e. There shall be at least 1 inch between clearance reduction systems and combustible walls and ceilings for reduction systems using ventilated air space.

f. If a wall protector is mounted on a single flat wall away from corners, adequate air circulation shall be permitted to be provided by leaving only the bottom and top edges or only the side and top edges open with at least a 1-inch air gap.

g. Mineral wool and glass fiber batts (blanket or board) shall have a minimum density of 8 pounds per cubic foot and a minimum melting point of 1,500°F.

h. Insulation material used as part of a clearance reduction system shall have a thermal conductivity of 1.0 Btu inch per square foot per hour °F or less. Insulation board shall be formed of noncombustible material.

i. There shall be at least 1 inch between the appliance and the protector. In no case shall the clearance between the appliance and the combustible surface be reduced below that allowed in this table.

j. All clearances and thicknesses are minimum; larger clearances and thicknesses are acceptable.

k. Listed single-wall connectors shall be permitted to be installed in accordance with the terms of their listing and the manufacturer's instructions.

l. For limitations on clearance reduction for solid-fuel-burning appliances see Section M1306.2.1.

As with other mechanical appliances, access is required for service. These types of requirements are typical for all of the various types of heating and cooling appliances addressed in the code.

Exhaust Systems: IRC Chapter 15

As the name implies, mechanical equipment that removes air from the dwelling is referred to as an exhaust system. All air removed by an exhaust system is required to be discharged to the outdoors and cannot be recirculated within the dwelling. Chapter 15 of the IRC addresses most types of exhaust systems including clothes dryer vents, cooking range and open-top broiler hoods, microwave units installed over ranges, and bathroom fans.

Here, the IRC provides code criteria including requirements for ducts used to distribute exhausted air to the exterior. Types of materials, sizes, and exhaust duct lengths are addressed in various sections. The IRC also establishes required exhaust rates for kitchen hoods and bathrooms in Table M1507.3 to ensure that fans will be adequate to remove moisture, odors, and smoke and to ensure that changes of air inside the space will occur rapidly (see Figure 14-7).

Duct Systems: IRC Chapter 16

Chapter 16 of the IRC addresses ducts used in heating, cooling, and ventilation systems. We know that ducts are air distribution systems from our previous discussions. Here, the IRC provides both the material standards and the methods of installation for ducts used in dwellings. Ducts are required to be designed as prescribed in ACCA Manual D, the standard for duct sizing and layout. The construction of above-ground ducts is regulated under the provisions of M1601.1.1, and the construction of below-ground ducts is regulated under the provisions of M1601.1.2. Above-ground duct systems are limited to those systems that discharge air at not greater than 250° F (121° C). Additionally, the IRC requires ducts to meet the criteria provided in the Sheet Metal and Air Conditioning Contractors National Association (SMACNA) Fibrous Glass Duct Construction Standards, or the North American Insulation Manufacturers Association (NAIMA) Fibrous Glass Duct Construction Standards.

This code chapter provides installation criteria as well, as it addresses the treatment of joints and seams in ducts, and their support. Duct insulation and vapor barrier requirements are found here, and access requirements and other related criteria. Duct systems for both **supply air** (the air that is distributed to condition the space) and **return air** (the air that is returned to the furnace for reconditioning) are

Figure 14-7

Table M1507.3 Minimum required exhaust rates for one- and two-family dwellings.

TABLE M1507.3
MINIMUM REQUIRED EXHAUST RATES FOR
ONE- AND TWO-FAMILY DWELLINGS

AREA TO BE VENTILATED	VENTILATION RATES
Kitchens	100 cfm intermittent or 25 cfm continuous
Bathrooms—Toilet Rooms	Mechanical exhaust capacity of 50 cfm intermittent or 20 cfm continuous

For SI: 1 cubic foot per minute = 0.4719 L/s.

Figure 14-8

A return air grille installed in the ceiling of a dwelling.

COURTESY DELMAR/CENGAGE LEARNING

TAKE NOTE!

Where electrical wiring is located within a plenum space, it must be *plenum rated* and specifically approved for plenum applications. Another option is the wiring may be installed in approved conduit or raceways.

addressed (see Figure 14-8). The IRC establishes important criteria for return air or outdoor air and prohibits return air from being taken from spaces where hazards exist, such as where flammable vapors are present or where the return air inlet is within 10 feet (3,048 mm) of vent opening, plumbing vent, or exhaust vent discharge. The idea being, of course, that it is undesirable to draw air into the dwelling from areas where odors, gas, or other hazards are present.

Section M1601.5 prohibits the use of under-floor areas as **plenums** (a chamber or space used as a part of the air distribution system) in new structures. This section allows modification and repairs to plenums in existing structures. The code generally prohibits plumbing waste cleanouts from being located in these areas, unless the plenum is located in an unvented crawl space that receives conditioned air per Section R408.3. Because existing plenums are a part of the air distribution system, all materials located within them, including sidewall insulation, must meet a flame-spread rating not greater than 200 as determined by ASTM E 84. Where under-floor areas are used as plenums, access must be provided for service.

Figure 14-9

A listed and labeled ductless range hood installed over a gas cooktop. Note the clearance to the combustible wood cabinet above to cooking surface.

COURTESY DELMAR/CENGAGE LEARNING

Special Fuel-Burning Equipment: IRC Chapter 19

To its credit, the IRC is also equipped to deal with other appliances and specialty equipment, some of which is not ordinarily encountered in the average dwelling. Fuel-burning ranges and ovens are addressed here, as are specialty items such as sauna heaters, stationary fuel cell power plants, and gaseous hydrogen systems.

Per the requirements of Section M1901.1, fuel-fired cooking ranges are required to have a vertical clearance above the cooking surface of 30 inches (762 mm) from unprotected combustible materials (such as kitchen cabinets). Other household cooking appliances are required to be listed, labeled, and installed per the manufacturer's installation criteria and must not be installed in such a manner that they will obstruct the required combustion air or prohibit their operation and service (see Figure 14-9).

Specialty items such as fuel-fired sauna heaters are also required to be installed per the manufacturer's specifications (a common theme in the mechanical world as we are starting to see). They must also be supplied with an adequate supply of **combustion air** (air used in the burning of a fuel). Other specialty mechanical systems include stationary fuel cell power plants and gaseous hydrogen systems, which will be addressed more fully in this and other chapters.

Boilers and Water Heaters: IRC Chapter 20

Both boilers and water heaters are addressed in this IRC chapter and the code provides criteria for each. While both types of units are used to heat water, it is important to understand that there are fundamental differences between these two types of water heating systems. Boilers are often more sophisticated and may involve special skills in their installation and inspection. Some states, such as Oregon, regulate boilers through state agencies instead of the local building department. Be sure to check with the jurisdiction where the proposed dwelling will be constructed for specific local requirements.

Per the definitions provided in the IRC, a "boiler is a self-contained device from which hot water is circulated for *heating* purposes and then returned to the boiler." Boilers operate at a pressure "not exceeding 160 pounds per square inch gage (psig)

Figure 14-10

A modern residential boiler installed in a dwelling. Boilers often require special skills and certifications or licenses for installation due to higher operating temperatures and pressures.

COURTESY OF JIM PERUSICH, JP HEATING & COOLING, LLC

(1103 kPa gauge) and at water temperatures not exceeding 250° F (121° C)." A water heater is a heating appliance or piece of equipment that heats water and supplies that water in a potable, hot water distribution system.

Water heaters generally supply hot water throughout a dwelling for potable use. They supply hot water for bathing, cooking, drinking, laundry, and similar applications. Hot water systems operate at temperatures and pressures well below those of boilers and by definition are part of the hot water supply. Hot water distribution systems are typically supplied with taps (faucets) to allow the homeowner access to the hot water supply. Although some hot water systems recirculate hot water (e.g., boilers), many do not and thus deliver water unidirectionally. In addition, some hot water distribution systems are used to supply hot water for both the potable and the space heating systems.

Boilers, on the other hand, are typically used for space heating (rather than heating water for potable use) and either heat water or generate steam in a *closed* system (see Figure 14-10). In other words, the homeowner would not typically open a tap to extract heated water from a boiler (for hand or clothes washing, for example) as one might do in systems that utilize a water heater.

The IRC requires both boilers and water heaters to be installed per the provisions of the chapter and per the manufacturer's installation instructions. Section M2002 of the IRC establishes important operating and safety controls for boilers due to their higher temperatures and pressures. Requirements are also established for expansion tanks, which are required for all boilers. Water heaters are regulated under the provisions of IRC Section M2005 and must also meet the code criteria established in IRC Chapter 24, Fuel Gas (if gas fired) and Chapter 28, Water Heaters (which addresses the plumbing code aspects of the installation) when installed in dual-purpose systems.

Hydronic Piping: IRC Chapter 21

Where hot water is also used for space heating purposes, it is distributed through hydronic piping. Hydronic piping can be used for boiler-supplied and water heater–supplied heating systems. All piping, valves, fittings, and other components in the system must be installed per the manufacturer's specifications and must be rated for use at the operating temperatures and pressures suitable for the hydronic system. Piping systems must be installed such that they can be drained and where they drain into the plumbing drainage system, the components must comply with the plumbing provisions found in IRC Chapters 25 through 33.

Hydronic piping must be properly supported and installed in such a way that it can resist excess stresses that might be induced through expansion, contraction, and building settlement. Piping must not be placed in contact with any other building materials that might cause corrosion or degradation of the pipe materials. Testing is required to ensure that the hydronic piping can safely and effectively operate without leaking. A **hydrostatic test** (tested with liquid water under pressure) is required at 100 psi (690 kPa) for a test period of not less than 15 minutes.

Hydronic piping is commonly used in floor heating systems (see Figure 14-11). Section M2103 of the IRC establishes material standards for these installations. Where the piping will be installed in concrete or gypsum floors, steel piping and copper tubing are common metallic materials used in these applications. A variety of plastic materials are used as well, including cross-linked polyethylene tubing (PEX) and chlorinated polyvinyl chloride (CPVC). Section M2103.3 of the IRC prescribes joint treatment requirements, which vary based on the types of materials used. Steel pipe joints are required to be welded; copper tubing joints are required to be **brazed** (similar to being soldered, but with a higher temperature metal filler material); and PEX tubing joints are required to be joined with compression, insert, or cold expansion fittings.

Figure 14-11

A hydronic or "radiant" floor piping system awaiting encasement in concrete.

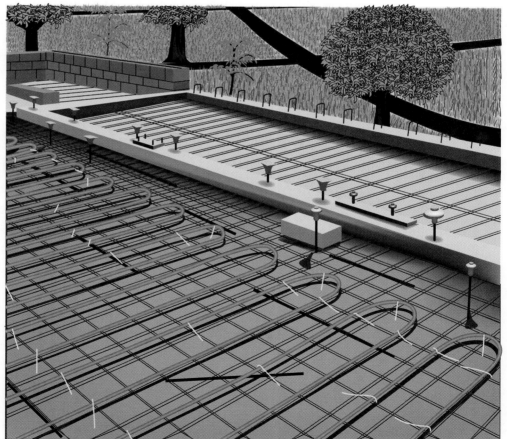

Ground-source heat pump loop piping is also addressed in this portion of the code. Materials used for ground-source systems must be approved for this specific application. Like hydronic piping, which will be embedded in concrete, ground-source loop piping must be pressure tested to 100 psi (690 kPa), but the minimum test period is increased to 30 minutes' duration. Trenches where such piping is installed may not be backfilled until the system is inspected and it has been determined that there are no observed leaks, nor are there any drops in pressure or flow rates that deviate from calculated values by more than 10%.

Solar Systems: IRC Chapter 23

Solar energy systems are regulated under multiple codes. The provisions found here are applicable to the mechanical portion of the system. The provisions of this IRC chapter are applicable to any solar energy system that uses solar energy for space heating or cooling, water heating, or swimming pool heating.

Solar energy systems must be installed so that all components are accessible for service, replacement, maintenance, and inspection. Solar collector panels are commonly installed on the roof of a dwelling (see Figure 14-12). Where this occurs, the roof must be designed to support the weight of the solar panels and related equipment. Supports for the panels must be noncombustible or constructed of fire-retardant-treated wood.

Solar systems can operate at extreme temperatures and pressures and must be provided with a pressure and temperature relief valve per IRC Section M2301.2.3. Solar systems must also be provided with a vacuum relief valve where subject to pressure drops below atmospheric pressure when operating or during shutdown. Solar panels must also be provided with freeze protection for the heat-transfer fluid where exposed to winter design temperatures below 32° F (0° C).

Figure 14-12

An array of solar panels mounted on the roof of a dwelling gathering free energy from the sun.

Solar systems must be installed so that the solar loop can be isolated by a valve for service and maintenance. Panel and other components must be listed and labeled for their intended use and must provide the manufacturer's identification information, temperature and pressure ratings for the panel and components, weight of panel, and so forth, to ensure that the system and supporting structural elements can be adequately designed. Systems must be equipped with a means to limit temperatures within the system to 180° F (82° C).

GREEN OPTIONS

Mechanical systems consume energy and it is for this reason that there is a great deal of interest in making them green. Energy consumed in an HVAC system is provided from either a fuel source or electricity. Fossil fuels, such as fuel oil, natural gas, and propane, are derived from products extracted from the earth and processed to make them useful as fuel.

It is likely that each person reading this text is well aware of the issues associated with fossil fuels, so we will not cover these issues at length. Suffice it to say that using less of these is a good thing. Thus, there is also a great deal of opportunity to make a dwelling greener by choosing alternative heating systems, by improving the performance of conventional heating systems, or by selecting high-efficiency systems to be used within the green dwelling.

Like so many areas in the code, opportunities to make mechanical systems greener are determined, in part, by one's perspective. Where the green decision is based on the belief that fossil fuels should be avoided, electrically operated heating systems will likely be the preferred choice. Therefore, traditional heat pumps, ground-source heat pumps, and other electrically operated heating and cooling systems might be best suited in these cases.

Where there is no aversion to the use of fossil fuels, high-efficiency gas furnaces might be the HVAC system of choice in the green dwelling. Additionally, a gas-fired boiler or water heater might be used to provide heated water for use in an efficient, hydronic heating system. Other choices exist as well, such as with alternative fuel sources like solar energy systems. The availability of a particular fuel type or lack thereof in certain regions also may limit choices, regardless of one's beliefs and desires. The IRC is well equipped to deal with such choices, as we will see in the coming sections.

IMPROVING HVAC PERFORMANCE

Even where conventional HVAC systems are chosen for use in the green dwelling, many improvements can be made through careful preplanning of the heating and cooling system and green decision making along the way. High-efficiency heating and cooling equipment is readily available and can be installed in virtually the same way and in the same amount of time as equipment of standard efficiency. Duct efficiency can be improved with relative ease through sealing, and heating or cooling systems can be split or "zoned" to save energy in unoccupied spaces. The following sections explore in detail these options and more.

High-Efficiency Furnaces and Heat Pumps

Although 40% or more of a dwelling's energy use is for space conditioning, it is interesting to note that the IRC makes no attempt in this or any chapter to prescribe a minimum level of performance (efficiency) for mechanical equipment. Although the IRC does not establish minimum standards, the federal government does, for both fuel-burning furnaces and all heat pumps. Some states adopt their own energy codes, including Oregon and California. Even though most states adopt standards well above federal guidelines, they are prohibited from allowing mechanical equipment with *lower* performance standards than the federal guidelines.

Current minimum federal standards for furnaces require that they have an Annual Fuel Utilization Efficiency (AFUE) rating of 0.78, or 78% efficiency. This represents a significant improvement over the furnaces of old, which were only 65% efficient. That's right—*35%* of the energy consumed in old furnaces was lost through the flue and other means! Because heat pumps have both a heating and a cooling cycle, federal guidelines require a minimum standard of performance for both cycles. Current federal guidelines for the heating cycle require a Heating Season Performance Factor (HSPF) rating of 7.7 and a Seasonal Energy Efficiency Ratio (SEER) for the cooling cycle of 13.

Green Decisions and Limitations

Shall I take steps to ensure that my conventional HVAC system is as efficient as possible?

How will the requirements of the IRC impact this decision?

Options

When selecting a furnace or heat pump to satisfy the requirements of IRC Parts V and VI, consider the following green options.

High-efficiency furnaces: A number of highly efficient, conventional heating systems are on the market today. Many of these furnaces carry the U.S. Environmental Protection Agency (EPA) Energy Star label. Energy Star furnaces must have an AFUE of at least 0.90, or 90% efficiency rating. An AFUE of 95% is not uncommon and costs for these super-efficient furnaces have become quite reasonable in many areas (see Figure 14-13). These furnaces perform more efficiently because of improvements such as secondary (condensing) heat exchangers, two-stage burners for efficient operation under both mild and cold winter conditions, and electronic (pilotless) ignition. Rebates and tax credits may also be available for these systems, creating an even stronger incentive to select high-efficiency models.

High-efficiency heat pumps: Like furnaces, a number of highly efficient heat pumps are readily available today. Many heat pumps also carry the U.S. EPA Energy Star label. To qualify for the Energy Star label, packaged air-source heat pumps must have a HSPF of 8 or greater and a SEER of 14 or greater. Split systems must have a HSPF

Why It's Green

Improving the efficiency of conventional HVAC systems wherever possible is green because significant energy savings can be realized when compared to systems of average efficiency. Highly energy-efficient heating and cooling systems are considered more sustainable because they use less fuel and produce less pollution than their conventional counterparts. Efficiency can be improved—sometimes significantly—through careful planning and by taking simple steps during construction to improve efficiency.

Figure 14-13

A 95% efficient gas furnace installed in a garage. Note the PVC pipes used to vent flue gas and to provide combustion air.

COURTESY DELMAR/CENGAGE LEARNING

of 8.5 or greater and a SEER of 14.5 or greater. Improvements in heat pumps are provided by improved heat exchangers, two-speed compressors, high-efficiency motors, and variable-speed fans. Other high performance heat pumps include both air-source and ground-source models, which can improve performance over traditional models by up to 30% and 75%, respectively.

Other efficiencies: Efficiencies can be gained in other areas as well. Some of these methods have been discussed in prior chapters, such as high-performance duct sealing and by keeping all ducts within the building thermal envelope. Other efficiency-improving measures include zonal heating and cooling, which allows heating and cooling only certain portions of the dwelling but not others depending on use and time of day. For example, basements that are used only occasionally might be located on a separate, independent system with their own thermostats. Other zonal systems allow rooms such as bedrooms (which are frequently empty during the day) to be kept at a cooler temperature than frequently used rooms such as the family room and kitchen, resulting in energy savings.

Alternative Heating Systems

Some heating systems offer alternatives to the traditional forced-air systems installed in the majority of homes today. Radiant heat systems and heating systems that utilize solar energy are but a few of these alternative systems. Radiant heat systems allow heat to radiate into the space by direct transfer from radiators, floors, and other means. Some of these systems derive at least some of the energy used from solar power, providing a big boost in efficiency in the form of free energy from the sun.

Alternative heating systems can improve indoor air quality, too, because they do not distribute air through ducts as do forced-air systems. In this way, they do not circulate dust, allergens, and other particulate matter that can cause indoor air pollution, another important factor in green construction. These systems may require a source of fresh air to maintain air quality. Alternative systems can be zoned as can their forced-air cousins, to ensure that heat is only provided where needed. Alternative heating systems can also be placed on timers to further reduce heating costs at certain times of the day when a dwelling is empty or when the owners are asleep.

Why It's Green

Alternative heating systems are green because they can improve indoor air quality and because they often incorporate free solar power for at least a portion of their operation. Systems that are electrically operated are considered by some to be more sustainable because they do not rely on fossil fuels and they help to reduce outdoor air pollution because they do not burn fossil fuels. Automatic controls help these systems to operate more efficiently by raising and lowering temperatures as needed throughout the day.

Green Decisions and Limitations

Shall I incorporate an alternative heating or cooling system in my green dwelling?

What IRC provisions apply?

Heating System Option

It is known that renowned architect Frank Lloyd Wright was enamored of Japanese culture and was an avid collector of Japanese prints, gleaned during his frequent visits to the country. Once, after being commissioned to design the Tokyo Imperial Hotel, he visited a nobleman whose home was heated in the Korean "ondol" or "warm stone" method (heated floors). Wright later referred to this experience as "the indescribable comfort of being warmed from below" and for many decades afterward it was the standard heating system in all his designs.

Solar hydronic heating creates the same comfortable form of heat within a dwelling and is prized for this reason. Additionally, radiant-type heating systems are sometimes credited with providing a more comfortable interior environment because they do not create extreme dry conditions as do forced-air systems. When combined with solar panels, hydronic heating systems can be incredibly efficient, as part or all of their heating energy can come from the sun.

Alternative Energy Sources

Just as there are numerous options for types of HVAC systems, there are also options for energy sources for the operation of the HVAC systems. New alternative energy options represent an exciting area for the green dwelling, because they provide clean alternatives to fossil fuels and move away from other traditional sources, such as coal and hydroelectric power. As we have already mentioned several times, solar power is an attractive option. After all, solar energy is free, abundant, and clean.

TAKE NOTE!

Solar-powered hydronic heating systems must not only meet the code criteria established in IRC Chapter 23, but also meet applicable code requirements from Part VII of the code, Plumbing, in IRC Chapters 25 through 33, where the system is also part of the plumbing or potable water system. These systems must also meet the applicable criteria in Part VIII of the code, Electrical, which contains IRC Chapters 34 through 43.

Other alternative forms of energy have become recent topics of interest. Wind, water, and geothermal energy sources are being explored with great interest. In some areas, the incorporation of such technology is possible. For example, in areas where wind is abundant (in mountainous areas or where waterways are prevalent), turbines can be utilized to provide some electrical energy. Where streams and creeks are prevalent, water turbines can be utilized with some success. Granted, some of these technologies are impractical, at least by today's standards, but as we have seen with other forms of technology, what seems to be out of reach today becomes standard practice tomorrow. For this reason, they warrant consideration here.

Green Decisions and Limitations

Shall I incorporate an alternative energy source into my green dwelling?

What limitations exist within the code?

Energy Source Option

Wind power represents an exciting new technology because it utilizes wind turbines to harness the power of the wind and convert it to usable electricity. Turbine power is particularly effective in areas where the natural terrain fosters the flow of wind. Obviously, wind-sheltered areas are not ideal for this type of energy, so a cost-to-benefit analysis must be performed prior to investing money to ensure that such a system will perform adequately. Wind power can be harvested regionally in large-scale, commercial operations and fed into the power grid or, with available new technology, it can be utilized on individual properties through the use of smaller turbines.

Why It's Green

Alternative forms of energy are considered more sustainable because it is widely recognized that reliance on fossil fuels must be reduced in the near future. Fossil fuels are less abundant than they were even a few decades ago and are widely recognized as chief contributors to air pollution and climate change. Alternative forms of energy utilize clean energy sources provided in nature with minimal impact to the environment.

Wind turbines are available as tower-mount and roof-mount types. As this technology has developed, turbines have become smaller and more efficient, making them easier to incorporate into the green construction project. A number of turbine kits are now available, as are completed turn-key systems that are ready for use once installed. Wind turbines most commonly consist of a blade propeller (similar to that on an airplane) but are also available in vertical axis models in helical configurations, drum-type models, and others (see Figure 14-14).

Wind turbines convert wind into electricity by turning a generator as they spin; thus, they provide a supply of electrical current that is fed into the building's wiring. This current can also be placed back into the power grid through the use of an inverter that must be carefully designed and installed to ensure a safe, code-compliant installation. Be sure to check with the local jurisdiction for all requirements prior to purchase and installation. The installation must meet all

Figure 14-14

A blade propeller
wind turbine.

COURTESY DELMAR/CENGAGE LEARNING

applicable provisions of Part VIII, Electrical, of the IRC. Whether roof or tower mounted, turbines and their supporting structures must be designed to resist all wind, snow, and seismic loads. Be sure to check with the local electrical utility provider to ensure that local requirements are met as well.

CHAPTER 14 APPLIED TO THE SAMPLE PROJECT

Chapter 14 pertains to mechanical systems used for the purpose of providing heating and cooling in one- and two-family dwellings. It provides general code criteria for the installation of furnaces, heat pumps, air conditioning, hydronic systems, and more. Additionally, Chapter 14 addresses ducts, controls, and other components of the HVAC system. The following analysis will demonstrate the application of some of the IRC provisions for mechanical systems with consideration given for the proposed scope of work and our design philosophy.

Site Considerations

Recall that the scope of work for the sample project included specifications for the use of a high-efficiency geothermal or "ground source" heat pump. As we have seen, the IRC provides various code criteria governing the installation of such systems. Ground-source heat pumps utilize water piping or tubing buried in the earth as a source of heat in heating months and as a heat sink during cooling months. As such, it will be necessary to excavate trenches to install the heat pump loop piping or tubing at the proper depth.

We will want to check with the local building department to determine if a grading and excavation permit is necessary for this work. Additionally, we will want to thoroughly check all local zoning ordinances and other regulations to ensure that the heat pump loop piping and related mechanical equipment will be located on the site in a manner that will meet setback requirements. The locations of the loop piping or tubing and all mechanical equipment should be accurately located on a site plan so that the proposed locations can be reviewed by the jurisdiction.

Building Considerations

For this project, we have chosen a ground-source heat pump that will utilize an air heat exchanger and blower motor to distribute heated or cooled air through the sample dwelling. The heat pump must meet the requirements of IRC Chapter 14 for access and sizing as well as all requirements identified in the manufacturer's installation instructions. In this case, the heat pump will be installed outdoors so it must be listed and labeled for outdoor installations. It must also be located on the site such that it will satisfy local ordinances for setbacks and noise, where applicable.

The ground-source loop piping or tubing must meet the general requirements of IRC Section M2101, the requirements of IRC Section M2104, Low Temperature Piping, and must be of an approved material. Ground-source loop piping or tubing must be pressure tested to 100 psi (690 kPa) for 30 minutes to ensure that it is free of observed leaks, per Section M2105.

Because the heat pump system will utilize forced air as the means of delivery of the heated or cooled air, it will be necessary to install ducts in the unvented attic of the dwelling. Recall that unvented attics help to improve duct system efficiency by keeping temperatures within the attic space at more moderate levels. All supply and return ducts must meet the requirements of Chapter 16, Ducts, and must be designed in accordance with ACCA Manual D.

CHAPTER SUMMARY

Chapter 14 of the IRC addresses the various code requirements for mechanical systems installed in dwellings. It provides standards for materials and equipment as well as requirements for construction methods, to ensure that HVAC installations are safe and suitable for their intended purpose. Following are highlights of the chapter.

- This chapter addresses requirements for residential mechanical systems from Parts V and VI of the IRC, which includes selected material from IRC Chapters 12 through 24.
- The mechanical code provisions address heating and cooling equipment, exhaust systems, ducts, specialized equipment, solar energy systems, and more.
- Some chapters are not addressed because they are not relevant to the discussion of green dwellings or because they offer no practical green options.
- Mechanical systems supply heated or cooled air to condition the space inside the building thermal envelope.
- Mechanical systems replace heat lost through the thermal envelope in winter months and remove heat that enters the dwelling during summer months.
- Heat flows from hot to cold through the building thermal envelope as long as there is a temperature differential between the two systems.
- Mechanical systems consume 40% or more of the energy used in a dwelling for the purpose of space conditioning and are thus great candidates for green improvements.
- The efficiency and performance of traditional mechanical systems can be improved by selecting high-efficiency models (commonly called Energy Star), sealing ducts, and using zonal systems.
- Alternative heating systems make excellent choices for green dwellings and many, such as hydronic heating systems, are allowed in the IRC.
- Alternative forms of energy, especially when combined with highly efficient or alternative heating and cooling systems, can be particularly effective at reducing energy consumption and increasing system efficiency.
- Manufacturer's installation guidelines for all equipment and components must be carefully observed.

PLUMBING SYSTEMS

15

This chapter corresponds to Part VII of the IRC, Plumbing, and includes selected material from IRC Chapters 25 through 33.

learning objective

To know and understand how plumbing systems are generally regulated by the code. Also, to explore green options for plumbing systems, fixtures, and related components and to understand how the IRC requirements impact those choices once made.

IRC CHAPTERS 25 THROUGH 33—OVERVIEW

Like the previous chapter on mechanical systems, this portion of the text will explore selected International Residential Code chapters related to the topic of residential plumbing. Collectively, all plumbing regulations that are relevant to the construction of a one- or two-family dwelling are presented in Part VII of the IRC, Plumbing. This "going green with the IRC" chapter will cover topics related to general plumbing requirements, plumbing fixtures, water heaters, and water supply systems and will explore green options and choices for the plumbing in our green dwelling.

In this chapter, we will use the term "plumbing" to reference piping systems in general and when it appears, it will generally include all components of the plumbing system from the fixtures to the water service and everything in between. Where specific plumbing components are addressed, they will be referred to by name or as "fixtures" and the text will make it clear what plumbing system component is being referenced.

Chapters 25 through 33 of the IRC are devoted to the topic of plumbing and storm drainage systems, which are not specifically addressed in other areas of the code (such as hydronic piping systems, for example, the requirements of which are found in the mechanical provisions of the IRC). Where the IRC does not contain specific code provisions to address a particular fixture, plumbing piping, or other system component (such as roof drain sizing), the reader is referred to the applicable provisions of the 2009 International Plumbing Code.

Modern plumbing systems are used to provide and maintain a sanitary environment within a dwelling. Plumbing systems deliver a supply of fresh, safe water that is used for cooking, drinking, bathing, and clothes washing (see Figure 15-1).

Figure 15-1

Plumbing piping for a typical laundry room. In this case, from left to right is the washer box with hot and cold water supply, the electric dryer location, and rough plumbing for a laundry tray (sink).

COURTESY DELMAR/CENGAGE LEARNING

TAKE NOTE!

Onsite waste disposal systems (i.e., septic tanks and drain fields) are beyond the scope of the IRC. Most states, counties, and local municipalities regulate the disposal of human waste on the dwelling site, where municipal sewage systems are not available. Onsite systems are complex and often regulated on multiple levels. Be sure to check all state and local laws and regulations *prior to the design and installation* of an onsite waste disposal system. As a resource, consult the 2009 edition of the International Private Sewage Disposal Code (IPSDC) for additional guidance. The complete version of the 2009 IPSDC is also included in the 2009 International Plumbing Code. Obtain all necessary permits prior to construction.

Safe drinking water is sometimes taken for granted in the United States because we are so accustomed to having a convenient, abundant supply at our fingertips. One need travel only as far as the kitchen sink for what appears to be a seemingly endless supply of fresh water. Modern plumbing systems installed under the plumbing provisions of the IRC and other recognized codes are to thank for this great convenience and privilege.

Plumbing systems also help to maintain a sanitary, safe indoor environment by carrying waste products from the dwelling and disposing of them in approved sewage disposal systems. Building drains carry human waste products out of the dwelling and dump the waste into the building sewer, on its way to a municipal sanitary sewer or to an approved onsite waste disposal system. In so doing, building drains and sewers that terminate in an approved manner greatly reduce the risk of disease-borne fecal coliforms, E. coli, and other pathogens that can potentially contaminate supplies of drinking water through improper disposal.

PLUMBING SYSTEMS AND THE IRC

Chapter 15 of *Going Green with the IRC* covers a range of topics in the code that collectively make up the requirements for plumbing systems. For clarity and to ensure a complete treatment of this topic in the space allotted, only relevant chapters are explored. The material that is necessary for consideration in the green dwelling is presented here; however, the code sections presented in this chapter are by no means a complete treatment of all code requirements. Consult the 2009 IRC for all applicable plumbing code provisions.

General Plumbing Requirements: IRC Chapter 26

Chapter 26 opens with a charging statement that draws all aspects of the residential plumbing system into the scope of its provisions. IRC Section P2601.2 requires all plumbing fixtures, drains, and appliances that are used to receive and discharge plumbing wastes and sewage to be connected to a sanitary drainage system. The intent of this section is to ensure that plumbing systems terminate in an approved manner. Where plumbing fixtures, drains, appliances, and water supply systems will be installed in flood hazard areas, they must be located and installed in accordance with IRC Sections R322.1.6 and R322.1.7 to ensure that they can resist flood forces and are generally flood resistant.

Where plumbing piping is installed or repaired, the work must be done in such a way that wood joists, walls, beams, and other structural components will be left in a safe condition. The IRC imposes restrictions and limitations on the placement, type and size of holes, borings, notches, and other modifications to ensure that such safe conditions will occur in both wood and steel framing. The removal of too much wood material or an ill-placed notch could potentially weaken a wood structural component to the point of failure under a load.

Likewise, where piping is installed in concealed locations through holes or notches in wood floor joists, studs and rafters, or similar members, the IRC imposes additional requirements to prevent unintended damage. Where such concealed piping is installed less than 1.5 inches (38 mm) from the edge of the wood member, steel shield plates (also called nail plates) are required to be installed on the edge of the wood framing member over the notch or hole to protect the piping from damage (see Figure 15-2). When piping is installed too close to the edge of a framing member, punctures from nails and screws are much more likely. Shield plates protect the piping and prevent inadvertent punctures and other damage.

The shield plates described previously provide one form of protection. The IRC also requires the protection of piping from breakage and corrosion per Section P2603.3 to prevent building movement from damaging pipes and to prevent corrosion or

Figure 15-2

Plumbing piping for a kitchen sink. Note the shield plates just below the window on the left side.

COURTESY DELMAR/CENGAGE LEARNING

chemical reactions from damaging the plumbing piping. Where piping passes through foundation walls or under footings, a relieving arch or a sleeved opening at least two pipe sizes greater than the pipe diameter must be provided. Annular spaces between the pipe and the sleeve must be filled to prevent the entry of water and insects. Fire ratings must be maintained where piping passes through fire-rated assemblies with approved materials such as fire-rated caulks or collars.

Piping must be protected under other installation conditions as well. The IRC prescribes the minimum depth requirements for both sewer and water supply piping to protect piping from freezing temperatures and to reduce the risk that the pip-

Figure 15-3

Figure P2604.4 Pipe locations with respect to footings.

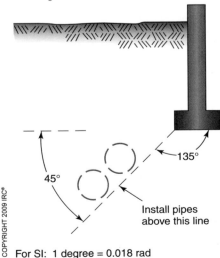

45°

135°

Install pipes above this line

COPYRIGHT 2009 IRC®

For SI: 1 degree = 0.018 rad

ing will be damaged from routine digging. The IRC also requires that such piping be located above the bearing plane of footings as shown in Figure 15-3. Piping must be supported along its entire length when installed in a trench or other excavation. Proper support ensures that piping will not overflex, which might potentially cause leaks at joints and material fatigue. Improper support also causes low spots or "bellies" that can obstruct flow in the pipe.

Most important, IRC Section P2608 establishes material standards for all plumbing piping, fittings, traps, fixtures, and appliances and requires strict accordance with manufacturer's installation criteria. It also requires compliance with the criteria established in the material listing. Some plumbing products and materials require either third-party testing or product certification by approved third-party agencies. Plastic pipe, fittings, and components must meet the requirements of National Sanitation Foundation (NSF) Standard 14 and must be third-party certified. Water supply pipes, fittings, valves, and similar must meet NSF 61.

Plumbing Fixtures: IRC Chapter 27

Plumbing fixtures of all sorts are addressed here and the IRC opens with minimum standards for fixture quality. The IRC requires plumbing fixtures to be constructed of approved materials, as it does with all code sections dealing with materials. Typically, plumbing fixtures are made of porcelain, plastic, porcelain-covered steel, and fiberglass. Plumbing fixtures must be smooth and free of defects. They must, of course, conform to all standards identified in the code, and all plumbing fixtures must be provided with an adequate supply of potable water to flush and keep the fixture clean and sanitary.

Fixtures must be securely fastened to their supporting structures and must be made water tight. It is very important that plumbing fixtures not be supported by the plumbing piping and fittings to avoid undue stress on these components. All plumbing fixtures must be installed with minimum clearances to ensure that they are usable (see Figure 5-13). Piping, fixtures, and equipment cannot be installed in such a way that they will interfere with the operation of doors and windows. Additionally, fixtures must be installed so that they and the piping and fittings to which they are connected can be serviced or repaired.

IRC Sections P2708 through P2724 establish minimum standards for the installation of plumbing fixtures typically encountered in a dwelling, such as showers, bathtubs, toilets, lavatories, and washing machines. In addition, minimum fixture sizes (as with showers), installation criteria, test procedures, and access requirements are established here. Given the limited space available, we will not explore the specific requirements for each of the different fixture types. Suffice it to say that the IRC provides comprehensive coverage for all fixture types and should be carefully followed to ensure a code-compliant installation.

Water Heaters: IRC Chapter 28

Water heaters are an interesting appliance from a code standpoint. They are unique in that they are regulated by multiple code sections—in fact, more than any other appliance. Recall from Chapter 14 of this text that water heaters are regulated under numerous mechanical provisions when used for space heating and where they burn a fuel such as fuel oil, propane, or gas to ensure they are properly vented. Water heaters are also regulated under provisions found in Part VIII of the IRC, Electrical, which provides code criteria for safe electrical connections and grounding. In areas prone to seismic activity, they must be properly supported and braced to the structure to ensure they will remain stationary during an earthquake, thus the structural provisions of the code come into play. Of course, they are covered under the plumbing provisions found here as well.

Thus, to ensure that they are installed in a code-compliant way, numerous sections of the code must be considered. From a plumbing standpoint, water heaters serve an important function because they provide a steady supply of hot water for sanitation and hygiene. IRC Section P2801.1 requires a water heater in every dwelling and must supply all fixtures used for bathing, washing, or cooking. Where water heaters are installed inside a dwelling or in areas where they can cause damage should a leak occur, a collection pan is required beneath the water heater or storage tank to collect water from the leak and distribute it to an approved drain or to the exterior of the building per IRC Section P2801.5.

Section P2802 of the IRC provides additional requirements where water heaters are used for space heating purposes. Piping and fittings used to supply water for space heating purposes must be approved for use in a potable water system. Furthermore, where the space heating system requires water to be heated to greater than 140° F (60° C) for its operation, a master mixing valve is required to mix the water to a temperature equal to or less than 140° F (60° C) for domestic use. This precaution is taken to prevent scalds from water that is too hot.

Water heaters are capable of generating dangerous temperatures and pressures should a malfunction of the built-in thermostat occur. For this reason, a separate temperature relief valve and a separate pressure relief valve or a combination temperature-pressure relief valve is required by IRC Section P2803 for every water heater, which will safely reduce temperatures and pressures should this unfortunate situation occur. These valves are required to operate within specific temperature and pressure ranges to ensure that conditions inside the water heater will not reach unsafe and potentially dangerous levels, per the provisions of Sections P2803.3 and P2803.4. A discharge pipe is required for each relief valve, to safely direct the discharge to the floor, thus averting personal injury or structural damage.

Water Supply and Distribution: IRC Chapter 29

A supply of potable water for every dwelling is one of the most important and fundamental requirements of the code. Potable water is used for a variety of household functions and is the most basic of modern conveniences. It stands to reason that the IRC would provide extensive code requirements on the topic of water distribution systems. These systems include not only the water supply piping but also valves, fittings, and other components that make up the water supply and distribution system.

Water is fundamental to human health and must be protected from contamination. Once contaminated, water can cause serious illness and lead to the spread of contagious diseases. Thus, the IRC imposes strict controls to prevent contamination from occurring. The potable water supply must be provided with some means to prevent **cross-connection** (an inadvertent connection between two otherwise separate piping systems) to eliminate the risk of contamination from harmful substances such as undesirable liquids, gases, or solids. Protection is also required from contaminates that could enter open piping systems such as lawn irrigation systems (e.g., pet waste or lawn chemicals).

Cross-connection protection is accomplished through special devices that provide directional control to the flow of water. Generally, this is referred to as backflow prevention. **Water-distribution backflow** is the flow of water or other liquids into the water supply system from other than the intended source. Backflow protection is addressed in IRC Section P2902.3 which establishes code requirements for different types of backflow prevention devices. Such devices include air gaps, reduced-pressure backflow prevention devices, and double check-valve assemblies, all of which have a specific function and application. Table P2902.3 identifies the specific application and types of devices acceptable under the code (see Figure 15-4).

Figure 15-4

Table P2902.3 Application for backflow preventers.

TABLE P2902.3
APPLICATION FOR BACKFLOW PREVENTERS

DEVICE	DEGREE OF HAZARD[a]	APPLICATION[b]	APPLICABLE STANDARDS
Air gap	High or low hazard	Backsiphonage or backpressure	ASME A112.1.2
Air gap fittings for use with plumbing fixtures, appliances and appurtenances	High or low hazard	Backsiphonage or backpressure	ASME A112.1.3
Antisiphon-type fill valves for gravity water closet flush tanks	High hazard	Backsiphonage only	ASSE 1002, CSA B125.3
Backflow preventer with intermediate atmospheric vents	Low hazard	Backpressure or backsiphonage Sizes ¼" – ¾"	ASSE 1012, CSA B64.3
Double check backflow prevention assembly and double check fire protection backflow prevention assembly	Low hazard	Backpressure or backsiphonage Sizes ⅜" – 16"	ASSE 1015, AWWA C510, CSA B64.5, CSA B64.5.1
Double check detector fire protection backflow prevention assemblies	Low hazard	Backpressure or backsiphonage (Fire sprinkler systems) Sizes 2" – 16"	ASSE 1048
Dual-check-valve-type backflow preventer	Low hazard	Backpressure or backsiphonage Sizes ¼" – 1"	ASSE 1024, CSA B64.6
Hose connection backflow preventer	High or low hazard	Low head backpressure, rated working pressure backpressure or backsiphonage Sizes ½" – 1"	ASSE 1052, CSA B64.2.1.1
Hose-connection vacuum breaker	High or low hazard	Low head backpressure or backsiphonage Sizes ½", ¾", 1"	ASSE 1011, CSA B64.2, CSA B64.2.1
Laboratory faucet backflow preventer	High or low hazard	Low head backpressure and backsiphonage	ASSE 1035, CSA B64.7
Pipe-applied atmospheric-type vacuum breaker	High or low hazard	Backsiphonage only Sizes ¼" – 4"	ASSE 1001, CSA B64.1.1
Pressure vacuum breaker assembly	High or low hazard	Backsiphonage only Sizes ½" – 2"	ASSE 1020, CSA B64.1.2
Reduced pressure detector fire protection backflow prevention assemblies	High or low hazard	Backsiphonage or backpressure (Fire sprinkler systems)	ASSE 1047
Reduced pressure principle backflow preventer and reduced pressure principle fire protection backflow preventer	High or low hazard	Backpressure or backsiphonage Sizes ⅜" – 16"	ASSE 1013, AWWA C511, CSA B64.4, CSA B64.4.1
Spillproof vacuum breaker	High or low hazard	Backsiphonage only Sizes ¼" – 2"	ASSE 1056
Vacuum breaker wall hydrants, frost-resistant, automatic draining type	High or low hazard	Low head backpressure or backsiphonage Sizes ¾" – 1"	ASSE 1019, CSA B64.2.2

For SI: 1 inch = 25.4 mm.

a. Low hazard—See Pollution (Section 202). High hazard—See Contamination (Section 202).

b. See Backpressure (Section 202). See Backpressure, Low Head (Section 202). See Backsiphonage (Section 202).

The IRC also requires protection of the potable water supply at the water outlets (faucets or hose bibbs, for example) to prevent backsiphonage. **Backsiphonage** occurs when used or contaminated water and other liquids are inadvertently drawn back into the potable water supply due to negative pressure in the system piping. Backsiphonage is prevented through the use of a variety of types of antisiphon devices and vacuum breakers. Such devices are required at fill valves, hose connections, and all other potable water openings and outlets. Provisions for the type and installation criteria are found in IRC Section P2902.4.

IRC Section P2903 provides prescriptive code criteria to ensure a safe, functional water supply system. Such systems must be adequately sized to provide a sufficient quantity of water at a suitable pressure for all water-consuming appliances installed within the dwelling. Thus, the water service must be designed for peak demand conditions so that the capacities at the outlet will not be lower than those established in Table P2903.1, as seen in Figure 15-5. Capacities are provided as flow rates in gallons per minute (gpm) at defined flow pressures in pounds per square inch (psi).

Plumbing fixtures are also subject to maximum flow rates in an effort to conserve water use. Table P2903.2 provides flow rates for residential plumbing fixtures that are consistent with the United States Energy Policy Act guidelines. Flow rate maximums have been in place since the early 1990s to reduce the amount of water used by Americans, especially for toilets, which account for more than half of the water used by U.S. residents. Under the federal guidelines, faucets for sinks and lavatories are limited to 2.2 gpm at 60 psi (8.3 L/m at 414 kPa), shower heads are limited to 2.5 gpm at 80 psi (9.5 L/m at 552 kPa), and toilets are limited to 1.6 gallons (6 L) of water per flush cycle.

System sizing information is provided as well. As mentioned, water distribution systems must be carefully sized to ensure the pipe sizes and available pressures during peak demand are capable of delivering the right amount of water for safe and effective operation of the plumbing fixtures (see Figure 15-6). Water systems are sized based on fixture units and flow rates in gallons per minute. The IRC provides convenient, easy-to-use prescriptive tables that make these calculations relatively easy to compute. Although not shown here for brevity, IRC Tables P2903.6, P2903.6 (1), and P2903.8 (1) provide the necessary information to establish fixture-unit values for plumbing fixtures, convert fixture-unit values into flow rates in gallons per minute, and to size manifolds, respectively.

As with all IRC sections, material standards are provided for water distribution piping, valves, fittings, and other components. Water piping can either be of metallic materials such

Figure 15-5

Table P2903.1 Required capacities at point of outlet discharge.

**TABLE P2903.1
REQUIRED CAPACITIES AT
POINT OF OUTLET DISCHARGE**

FIXTURE AT POINT OF OUTLET	FLOW RATE (gpm)	FLOW PRESSURE (psi)
Bathtub, pressure-balanced or thermostatic mixing valve	4	20
Bidet, thermostatic mixing	2	20
Dishwasher	2.75	8
Laundry tub	4	8
Lavatory	2	8
Shower, pressure-balancing or thermostatic mixing valve	3	20
Shower, temperature controlled	3	20
Sillcock, hose bibb	5	8
Sink	2.5	8
Water closet, flushometer tank	1.6	20
Water closet, tank, close coupled	3	20
Water closet, tank, one-piece	6	20

For SI: 1 gallon per minute = 3.785 L/m,
         1 pound per square inch = 6.895 kPa.

Figure 15-6

Plastic water distribution piping configured into a manifold (Zurn Quickport shown). The red piping is used for hot water and the blue is used for cold water.

as copper, brass, or steel or it can be of plastic materials such as cross-linked poly-ethylene tubing (PEX), chlorinated polyvinyl chloride (CPVC), and polypropylene plastic (PP) tubing and piping. Water service and distribution piping must comply with myriad material standards, largely established by ASTM but also by the American Water Works Association (AWWA), the American Society of Mechanical Engineers (ASME), and others. Specific methods of installation are prescribed and test methods for the various types of materials are spelled out in great detail as well.

Sanitary Drainage: IRC Chapter 30

Drainage, waste, and vent (DWV) piping comprises the drainage and ventilation system piping used in modern dwellings. DWV piping must be adequately sized so that the **building drain,** that portion of the piping contained within the building to a point 30 inches (762 mm) outside the dwelling, will be able to receive and distribute waste and sewage at peak demands and carry it to the building sewer. DWV piping is sized in a similar fashion to that of water system piping, by fixture units. The IRC provides Table P3004.1, which conveniently allows for prescriptive installation of a safe, effective drainage system.

Of particular importance in sanitary drainage systems is joint treatment. These treatments vary considerably based on materials selected and the manner in which they are used. Although metallic pipe is allowed in dwellings, the vast majority of DWV piping used in dwellings today is plastic, with black ABS piping being the most commonly used. Solvent cement adhesives used to join all types of plastic pipe must be specifically approved for the application at hand and these too must meet strict standards for quality. Solvent cements and primers are typically color coded (CPVC solvent cement is yellow, for example), which make them relatively easy to inspect after the joints are glued.

TAKE NOTE!

Solvent cement used to join pipes must be approved for the specific type of pipe it will join. Solvent cements effectively "weld" plastic materials together; that is, the solvents soften and fuse the plastic material to form a permanent bond. Solvents must be compatible with the piping materials to ensure a water-tight, permanent bond. Use these cements in well-ventilated areas and avoid contact with skin, eyes, and other sensitive areas.

Figure 15-7

A close-up of Type ABS drain piping. Note the ASTM grade markings on the pipe wall.

COURTESY DELMAR/CENGAGE LEARNING

Like water distribution system materials, DWV piping must meet specific standards for quality, again largely established by ASTM standards but also including AWWA and ASME standards (see Figure 15-7). Building drains and sewers must be provided with **cleanouts** (accessible openings in the DWV piping system used for the removal of obstructions) so that they can be serviced if necessary. The IRC identifies spacing requirements for cleanouts and requires cleanouts to be provided at changes of direction greater than 45 degrees (0.79 rad). The IRC requires cleanouts to be accessible, with a minimum of 18 inches (457 mm) clearance in front of cleanouts used to service pipes 3 inches (76 mm) and larger and 12 inches (305 mm) clearance in front of those used to service smaller pipes.

GREEN OPTIONS

Water is a precious and increasingly scarce commodity. In states where rainfall is scarce, such as those with desert climates, the need to conserve water is ever-present. Even in states where rainfall is abundant, such as in the northwestern United States, water is not to be squandered as periodic drought conditions can affect these areas as well. It goes without saying that the effects of climate change are having an effect on rainfall, groundwater, and wetlands in many areas of the world. Water conservation, then, is paramount when discussing sustainability. Steps taken to conserve and protect our water resources benefit everyone.

With respect to plumbing systems, water conservation is the name of the game. Reducing the amount of water used in the operation of various fixtures and appliances has direct benefits not only for the homeowner but for the water purveyors as well, because reduced demands equate to lower operating costs. In many cases, reduced demands for water also mean less energy is needed to operate equipment, another plus for water conservation. The following sections will explore in detail a number of green options that exist for green plumbing systems.

IMPROVING PLUMBING SYSTEM PERFORMANCE

Plumbing systems can be improved in a number of ways. One of the simplest and most effective ways to improve system performance and increase efficiency is to use less water. Where the green dwelling is supplied by a municipal system or other water purveyor, there is an immediate and obvious advantage in this approach: Use less, then pay less. Even where the water supply is provided by a well, there are advantages to using less water. Less energy is needed to operate the well's pump equipment, thus reducing wear and tear on the equipment. Plumbing system performance and efficiency can also be improved by installing fixtures and appliances that consume less energy or that take advantage of the water recycling methods that exist today. The following sections will explore a few of these options.

Water-Conserving Fixtures and Appliances

When one says "low-flow" plumbing fixtures to the average person, the likely reaction is a mix of strong emotions ranging from exuberance to frustration depending on one's experience with such fixtures. For many, the very idea of low-flow fixtures conjures images of toilets that clog easily or at the very least require multiple flushes to clear the bowl completely. For others, it conjures images of shower heads that deliver a spray of water so weak that what would otherwise take 10 minutes becomes an hour-long affair. Of course, there have been "improvements" in low-flow shower head designs, if the dreaded pins-and-needle spray heads can be called an improvement! None is a particularly attractive option for the green dwelling, however.

Although the IRC is aligned with current federal requirements for maximum flow rates, the current thinking on water consumption is that more can be done. As technological advances have improved the quality and performance of low-flow fixtures, a number of options have developed that make attractive choices for the green dwelling. New, ultra-low-flow toilets and other high-performance fixtures are available that reduce water consumption significantly below the maximums established in federal rules; and the performance of low-flow fixtures has been greatly improved, making them effective and convenient.

Why It's Green

Water-conserving fixtures and appliances reduce the demand for and consumption of water, which is generally regarded as a sustainable practice. Potable water is high in embodied energy due to the processing needed to make it safe and the installation of piping systems to deliver it to dwellings, commercial buildings, and industrial plants. Lower demand for potable water means less energy needed to process it. Less water into the appliance means less discharge as well, reducing the amount of gray and black water waste that requires treatment. Reductions in demand for potable water reduce the cost of operating the dwelling—another bonus.

Green Decisions and Limitations

Shall I incorporate water-conserving plumbing fixtures and appliances into my green dwelling?

How do the provisions of the IRC affect this decision?

Options

When selecting plumbing fixtures and equipment that will meet the requirements of IRC Part VII, consider the following green options.

Water-conserving fixtures: In the last few years alone, numerous high-efficiency plumbing fixtures have emerged which make great choices for the green dwelling. Dual-flush toilets now give the user a choice between a half-flush for liquid waste products and a full-flush for solid waste (see Figure 15-8). Some models have two separate flush controls and others use single flush control with multiple stages to trigger the type of flush required. Such toilets save considerable amounts of water because they use only the amount of water necessary to clear the bowl.

Special aerators are available for faucets that discharge a satisfying, effective stream but use less water than their traditional counterparts. Some faucets are electronically activated by a sensor (often called touchless faucets) that provides water only when the hands or other object is placed in front of the activator. Because they run only when needed, touchless faucets conserve water. These types of faucets have the added advantage of being more sanitary because it is not necessary to touch a handle to activate it.

Eco-shower heads are now available that deliver satisfying streams of water with a full-flow feel but use less water than standard low-flow heads. Super-low-flow toilets are also available that use as little as 1.28 gallons (4.9 L) per flush cycle, yet clear the bowl as well as some traditional models that use three times as much water. Some of the high-performance toilets are of a flapperless design; that is, they utilize a stream of water under pressure to clear the bowl in lieu of a traditional flapper. In so doing, they effectively clear the bowl and eliminate the necessity for multiple flushes.

High-performance plumbing fixtures are readily available in most areas. They are installed in much the same way as traditional fixtures. It is essential that any plumbing fixture chosen for use in the green dwelling meet the materials and installation standards established in the IRC. When considering such fixtures, it is a good idea to discuss the installation with the building official or his representative *prior* to purchase or installation. The building official will want to know to what standards the fixtures were manufactured and tested as well as the specific installation criteria established by the manufacturer.

Instantaneous water heaters: Traditional water heaters heat a supply of water and store it in an insulated tank in 40, 50, and even 80 gallon (151, 189, and 303 L) capacities. As hot water is drawn from the tank, it is replaced by cold water which must then be heated. Water that has been heated to a predetermined temperature must also be reheated periodically as heat is lost through the walls

Figure 15-8

Controls on a dual-flush toilet. Note the one-half and full-flush icons that allow the user to choose which is appropriate.

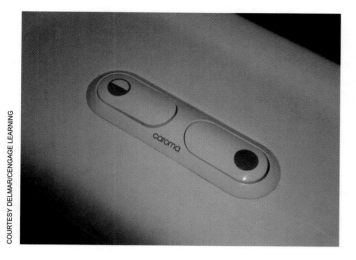

Figure 15-9

A gas-fired, tankless (or instantaneous) water heating system. Note how small this unit is compared to a traditional water heating tank.

of the tank, creating a never-ending cycle of energy use. Depending on the temperature set at the tank thermostat, more or less hot water may be needed at the shower valve to create a comfortable temperature.

Tankless or instantaneous water heaters heat water on demand rather than in the traditional way. Water is superheated in the tankless style water heaters, which means less of it is needed when combined at the mixing valve. Only the water needed at that moment is heated, giving instantaneous water heaters the added advantage of using less energy overall when compared to the traditional tank-type water heaters. This makes the choice to use the tankless models even more attractive.

Instantaneous water heaters are available in whole-house models that provide on-demand hot water to all fixtures within the dwelling (see Figure 15-9). Individual on-demand water heaters are also available that can be installed under a sink or adjacent to a shower. These models deliver smaller quantities of hot water but do so to a specific fixture. Instantaneous water heaters must be of an approved type for use in residential applications and must be installed in accordance with the terms of their listings. On-demand water heaters make a great solution for remote plumbing fixtures because they eliminate the need to run the water until the hot water makes its way from the tank to the fixture. This can save hundreds of gallons of water per year!

Gray Water Recycling Systems

While it may seem odd to describe water as "gray," in the plumbing industry it is common to describe the water discharged from certain plumbing fixtures and appliances as "gray," whereas water discharged from other fixtures is described as "black." Gray water, as it is defined in Appendix O of the IRC, is water that is discharged from lavatories, bathtubs, showers, clothes washers, and laundry trays. Black water is discharged from fixtures not included in the definition of gray water such as toilets, dishwashers, and kitchen sinks. The key difference is that black water contains human or food waste products. Gray water is typically discharged from fixtures used for washing and might contain soap residue and other by-products of the washing process.

Gray water recycling systems are not addressed in the general plumbing provisions in the IRC. Rather, these systems are addressed in IRC Appendix O (or Appendix C of the IPC), which must be specifically adopted by the jurisdiction if it chooses to make the provisions of the appendix chapter applicable within its scope of administrative authority. Section R102.5 of the IRC is clear that appendix chapters are not adopted unless specifically referenced in the adopting ordinance for the jurisdiction. Thus, unless a city or county adopts them, they cannot be construed to be enforceable or to establish mandatory requirements.

remember

In an earlier chapter we discussed the duties and responsibilities of the code official. He is directed to enforce all applicable provisions of the IRC. It is equally important to note, however, that he has authority to enforce only those provisions that are specifically adopted. If the appendices are not specifically referenced in the jurisdiction's adopting ordinance, then the building official has no authority to enforce their provisions.

What does it mean if a jurisdiction adopts Appendix O to govern the installation of a gray water recycling system? Appendix O, if adopted, has the effect of changing the text of IRC Section P2601.2 by inserting an important exception to the basic requirements of the code. Recall that P2601.2 requires every fixture to have a direct connection to the sanitary drainage system of the structure. Appendix O makes an exception for lavatories, bathtubs, showers, clothes washers, and laundry trays by allowing them to instead be connected to an approved gray water recycling system.

Gray water recycling systems can be used for the flushing of toilets and urinals (although admittedly, urinals are not frequently encountered within a dwelling) and also for subsurface landscape irrigation (see Figure 15-10). In either case, Appendix O of the IRC provides important criteria for the installation of these systems. Appendix O provides general criteria that are applicable to all gray water recycling systems as well as specific requirements that are applicable depending on which type of system is installed. Gray water recycling systems are *not* potable systems and must not be construed as such.

Generally, these systems must be constructed of materials that are found to be suitable under the provisions of the IRC. Above-ground DWV piping must meet the criteria outlined in Table P3002.1(1). Below-ground drainage and vent piping must meet the materials standards outlined in IRC Table P3002.1(2). Permits, inspections, and system tests are required, as for any plumbing system.

Gray water systems require a supply of water to be stored in a collection reservoir sized to accommodate twice the volume of water required to meet the daily flushing

Figure 15-10

A gray water recycling system collection tank; capacity 132 gallons (500 L).

requirements for all fixtures supplied with gray water. Connections between the potable system and the gray water recycling system may not be made, with the exception of the connection required by IRC Section AO102.3 which requires a supply of makeup water for systems intended for flushing toilets and urinals. Systems used for recycling water must be connected to a filtration system consisting of a media, sand, or diatomaceous earth filter. Collection reservoirs used for these systems must be provided with an overflow, venting, and rains for service and proper function.

Green Decisions and Limitations

Shall I use a gray water recycling system in my green dwelling?

How will the requirements of IRC Appendix O affect this decision?

Options

Where gray water recycling is desired and allowed by the code official, consider the following green alternatives.

Flushing of water closets and urinals: Section AO102 addresses gray water recycling systems used for the purpose of flushing toilets. In addition to the general requirements noted in the previous section, these systems require disinfection from an approved source, such as chlorine, iodine, or ozone. Makeup water is required to ensure that the system has the necessary supply of water to properly flush and clean toilets. Makeup water is provided through a connection to the potable system and is separated by an approved backflow device conforming to the requirements of IRC Section P2902.

Water distribution piping used for the recycling system must meet the requirements of IRC Table P2905.4. IRC Section AO102.4 requires water used in recycling systems to be treated with a food-grade dye, blue or green in color, before being supplied to the fixture. This coloring system makes it easy to see and recognize that recycled water is being used to cleanse the fixtures. Gray water recycling systems must be properly identified per Section 608.8 of the International Plumbing Code.

Subsurface landscape irrigation: Section AO103 of the appendix chapter establishes code criteria for the use of gray water for irrigation. Where such systems will be used, they are exempt from the makeup water, disinfection, and coloring requirements identified for systems used to flush toilets. This makes sense

Why It's Green

Gray water recycling is green because it allows water to be used more than once before being discharged into the sanitary sewer system. This is an inherently sustainable practice because it reduces not only the demand for clean, potable water but also the discharge that would otherwise occur with conventional systems. Additionally, gray water systems can be used for subsurface irrigation outdoors, which also reduces demands to the potable system.

because the water will be used below grade outside the home, which reduces the risk of contamination from pathogens or accidental consumption.

The subsurface irrigation system must be properly sized to ensure that the collection reservoir will be of sufficient size to accommodate the discharged water (see Figure 15-11). To ensure that soils are adequate to receive the discharge from the system, the permeability of the soils within the area of the proposed absorption system must be evaluated. It stands to reason that soils that do not drain quickly may not be well suited to this application. If the soils in question cannot accommodate the system, it could easily become overloaded and cause backups or other malfunctions. The IRC requires at least three soil percolation tests to be performed to determine the adequacy of the drainage system in each system area.

Subsurface irrigation systems that utilize recycled water must be located away from property lines, buildings, public water mains, seepage pits, wells, and other site features. Table AO103.8 provides minimum clearances from these and other site features. Thus, it is critically important to make the decision to utilize a gray water recycling system for irrigation purposes early in the design stages so that the site can be laid out accordingly.

Rainwater Harvesting Systems

Rainwater harvesting systems have gained popularity in recent years and are excellent candidates for use in the green dwelling. As the name implies, these systems are used to collect and distribute rainwater for later use. Such systems typically utilize a system of storage tanks and water distribution piping to route the harvested water to the fixture, appliance, or system where it will ultimately be used.

Rainwater harvesting systems use "catchment" tanks located either above ground or below ground to store the water (see Figure 15-12). Tanks may be steel,

Figure 15-11

A collection tank for a subsurface irrigation system.

COURTESY OF MARK MECKLEM, MIRANDA HOMES

Figure 15-12

A rainwater catchment tank (or rain barrel), capacity 60 gallons (227 L). Note the hose connection in the lower center portion of the tank.

COURTESY OF THE GREAT AMERICAN RAIN BARREL COMPANY

concrete, or plastic and are often quite large to accommodate what can amount to a sizable amount of water. The advantage is that water can be safely stored in climates where rainfall is scarce (desert climates) or where rainfall fluctuates significantly based on seasonal climatic changes.

Neither the International Residential Code nor the International Plumbing Code currently contains prescriptive provisions for the use of rainwater harvesting systems. Rainwater harvesting systems can certainly be proposed as an AMM, however. Some states, such as Oregon, have adopted prescriptive codes for rainwater harvesting that may serve not only as a useful resource but also as the basis for an initial AMM proposal to another jurisdiction. Consider discussing this option with the code official if plans for your project include a rainwater harvesting system.

Green Decisions and Limitations

Shall I incorporate a rainwater harvesting system into the design of my green dwelling?

How will the IRC requirements impact this decision?

Why It's Green

Rainwater harvesting is green because it allows water that falls in the form of rain to be collected and saved for use at a later time when water resources are scarcer. Rainwater is essentially free and reduces demands on the municipal system or private well. Water that would otherwise be shed from the building and disposed of on the property or in a storm sewer can be put to practical use at a later time.

Options

Where rainwater harvesting is desired and allowed by the code official, consider the following green options.

Rainwater used for irrigation systems: Rainwater can be used for irrigation purposes just as recycled gray water can. Rainwater is typically collected in a cistern during times when rainfall is abundant which can then be used for plants and lawns at a later time. Such systems directly reduce the amount of water that must be extracted from a municipal system to service landscaping, which saves money for the property owner as well as precious resources. These systems may also make irrigation possible for sites where a private well is not adequate to meet the needs of both the potable system and the demand for irrigation.

Rainwater used for drinking water systems: It is also possible to utilize rainwater for potable uses. Such systems harvest rainwater similarly to that used for irrigation purposes but then use sophisticated water treatment systems to make the water ready for potable use. Such systems use advanced filtration and disinfection methods to prepare the water for drinking. In effect, these systems provide a mini municipal water treatment facility on the dwelling site. Water distribution piping used in these systems is typically color coded to identify it from piping used for the ordinary supply to the dwelling (see Figure 15-13).

Figure 15-13

Piping used for reclaimed (recycled) water alongside copper water piping. Piping used in reclaimed water systems is always purple in color.

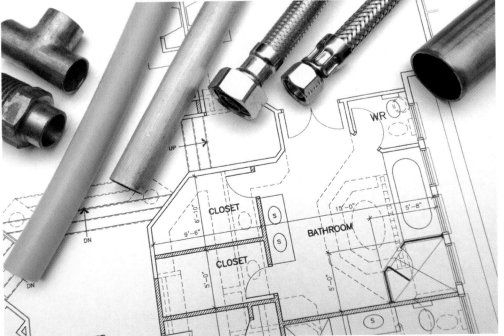

CHAPTER 15 APPLIED TO THE SAMPLE PROJECT

Chapter 15 pertains to plumbing systems used to deliver potable water to and distribute water throughout one- and two-family dwellings. It also provides code criteria for drainage, waste, and vent systems. Chapter 15 provides general code criteria for the installation of plumbing fixtures and water heaters. Additionally, Chapter 15 addresses storm drainage and other components of the plumbing system. The following analysis will demonstrate the application of some of the IRC provisions for mechanical systems with consideration given for the proposed scope of work and our design philosophy.

Site Considerations

For this sample project, we will assume that both a public sewer and water service are available to serve the proposed dwelling. Both the water supply piping and building sewer will require trenching to ensure burial at the depths identified in the IRC. The location and depth of available utilities must be identified and the installation of the plumbing systems must be carefully coordinated with the installation of the ground-source heat pump loop piping addressed in the previous chapter to avoid damage to this system.

The local building department, water purveyor, and sewer provider should be contacted for **as-built drawings** (drawings and other data that identify the specific locations of the installed utilities). We will want to obtain all necessary permits to enable the connection of water and sewer piping prior to performing any work.

Building Considerations

The scope of work for the sample project specified ultra-low-flow fixtures throughout the dwelling. A number of ultra-low-flow faucets and shower heads are readily available for residential sinks, lavatories, and showers. Many major manufacturers offer kitchen faucets with adjustable flow rates using a trigger or button-type adjusting mechanism. Kitchen and bathroom faucets and shower heads are available with flow rates as low as 1.5 gpm at 60 psi (5.7 L/m at 414 kPa). These types of faucets and shower heads fit both our specifications and scope of work nicely and will help to make our sample project as energy efficient—and water efficient—as possible.

We will also select dual-flush toilets that allow the user to select a full or partial flush depending on need. Toilets, or water closets as they are referred to in the IRC, must meet the maximum flow rate established in Table P2903.2, which is 1.6 gallons (6 L) per flush cycle. A number of ultra-low-flow toilets with flow rates as low as 1.28 gallons (4.9 L) per flush cycle are readily available; and we should verify that the flow rate for the full-flush mode in the brand and model we select does not exceed this rate. Both the water closet and the water closet tank must conform to one of the standards identified in IRC Section P2712.1, and the installation must meet all of the requirements of Section P2712.

Finally, our project specifications call for the use of a centralized, instantaneous or "tankless" water heater. Water heaters must be installed in accordance with the requirements of IRC Chapter 28, which covers the plumbing aspects of the installation, and must also meet the requirements of Chapters 20 and 24 from Parts V and VI, Mechanical. Because they are required to be installed per the manufacturer's installation instructions, we must familiarize ourselves with these requirements to ensure a safe, code-compliant installation. Domestic electric water heaters must conform to UL 174 or UL 1453, so we will want to make sure that the model we select meets this standard.

CHAPTER SUMMARY

Chapter 15 of the IRC addresses the various code requirements for plumbing systems installed in dwellings. It provides standards for materials and equipment as well as requirements for construction methods to ensure that plumbing fixtures, appliances, and systems are safe and suitable for their intended purpose. Following are highlights of the chapter.

- This chapter addresses requirements for residential plumbing systems from Part VII of the IRC, which includes selected material from IRC Chapters 25 through 33.
- Plumbing systems supply potable water and drain liquid waste products and sewage from inside the building envelope.
- The plumbing code provisions address plumbing fixtures, water heaters, water distribution systems, sanitary drainage systems, and more.
- Some chapters are not addressed because they are not relevant to the discussion of green dwellings or because they offer no practical green options.
- The efficiency and performance of traditional plumbing systems can be improved by selecting water-conserving fixtures and appliances (commonly called low-flow fixtures).
- Gray water recycling systems make excellent choices for green dwellings and are allowed in Appendix O of the IRC.
- The manufacturer's installation guidelines for all fixtures, appliances, equipment, and components must be carefully observed.

ELECTRICAL SYSTEMS

This chapter corresponds to Part VIII of the IRC, Electrical, and includes selected material from IRC Chapters 34 through 43.

learning objective

To know and understand how electrical systems are generally regulated by the code. Also, to explore green options for electrical systems, luminaires, and related components and to understand how the IRC requirements impact those choices once made.

IRC CHAPTERS 34 THROUGH 43—OVERVIEW

This portion of the text will explore selected IRC chapters related to the topic of residential electrical systems, in a similar fashion to that of the previous two chapters on mechanical and plumbing systems. Collectively, all regulations that are relevant to the installation of an electrical system in a one- or two-family dwelling are presented in Part VIII of the IRC, Electrical. This "going green with the IRC" chapter will cover topics related to general electrical requirements, electrical services, circuits, electrical equipment, and **luminaires** (lighting fixtures) and will explore green options and choices for the electrical systems in our green dwelling.

In this chapter, we will use the term "electrical" to reference wiring systems in general, and when it appears it will generally include all components of the electrical system from the service and circuits to the power distribution and everything in between. Where specific electrical components are addressed, they will be referred to by name and the text will make it clear what electrical system component is being referenced for clarity.

Chapters 34 through 43 of the IRC are devoted to the topic of electrical systems that might typically be encountered in one- and two-family dwellings. These IRC chapters are based on the National Fire Protection Association (NFPA) publication, entitled the *2008 Edition of the National Electrical Code* (NEC), and are used pursuant to license with the NFPA, copyright 2007 National Fire Protection Association, all rights reserved. Chapters found in the NEC are generally referred to as "articles" and the NEC provides electrical code criteria not only for dwellings but also for other types of buildings up to and including the largest, most complex industrial buildings. The provisions that are specifically applicable to dwellings are compiled into Chapters 34 through 43 of the IRC for convenience.

The electrical provisions found in the IRC are meant to address those wiring methods, systems, equipment, and related topics for electrical single-phase systems up to 120/240 volts, 0 to 400 amperes. Where the electrical service provided within the dwelling exceeds these limitations or where IRC does not contain specific code provisions to address a particular appliance, wiring method, or other system component (such as fuel cells), the reader is referred to the applicable provisions of the 2008 NEC. Additionally, the IRC allows wiring methods, materials, and subject matter covered in NFPA 70 to be substituted for IRC Chapters 34 through 43 and is considered equivalent to the requirements of Part VIII, Electrical, of the IRC.

Modern electrical systems are used to deliver a safe, dependable supply of electrical energy to the dwelling (see Figure 16-1). It goes without saying that the modern dwelling is totally dependent on electricity. One need only think back to the last time the power went out at home for a refresher on how true this is. Even with short-term outages, most of us are utterly lost without electrical power. Electrical wiring systems deliver a supply of electricity that is used to power appliances, operate lighting and outlets, and cool or heat our homes. Modern electrical systems installed under the electrical provisions of the IRC are to thank for this great convenience and privilege.

It is interesting to note that, while the IRC covers matters related to electrical safety extensively, the IRC does not directly address energy consumption or energy management. These issues fall outside the scope of the code and are left to other chapters such as IRC Chapter 11, Energy Efficiency. Still, it is important to understand the basic scope and intent of the electrical provisions as they will be presented in the following sections. Additionally, we will explore green options as they relate to electrical systems even though they are generally beyond the scope of the electrical code.

Figure 16-1

A modern electrical panel and service. Note the routing of the wires through the top of the panel and the connection to ground through the bottom of the panel.

COURTESY OF MARK MECKLEM, MIRANDA HOMES

ELECTRICAL SYSTEMS AND THE IRC

Chapter 16 of *Going Green with the IRC* covers a range of topics in the IRC that collectively make up the code requirements for electrical systems. For clarity and to ensure a complete treatment of this topic in the space allotted, only relevant chapters are explored. The material that is necessary for consideration in the green dwelling is presented here; however, the code sections presented in this chapter are by no means a complete treatment of all of the code requirements. Consult the 2009 edition of the International Residential Code or the 2008 edition of the National Electrical Code for all applicable electrical code provisions.

General Requirements: IRC Chapter 34

IRC Chapter 34 begins with the scoping provisions and charging statements as do other IRC chapters. IRC Section E3401.1, Applicability, establishes that the provisions of Part VIII, Electrical, apply to all one- and two-family dwellings and encompass the general wiring and systems that would typically be encountered in these residential structures. Section E3401.2 provides the scope for the electrical chapters in the IRC and references NFPA 70 as an approved alternate to the provisions established in the IRC. The language provided here makes it clear that an omission in the IRC electrical chapters of any material or construction method referenced in NFPA 70 shall not be construed as a prohibition to using such materials or construction methods. In simpler terms, if it is allowed by NFPA 70, it is acceptable under the IRC.

Like most building systems, the process of constructing the wiring system necessarily requires drilling, notching, and boring of structural components to facilitate the installation of wiring, electrical boxes, and related equipment. IRC Section E3402, Building Structure Protection, requires such modifications to wood framing members to conform to all applicable provisions established previously in the code. Where it is necessary to penetrate fire-rated assemblies (walls, floors, and ceilings), electrical installations must be adequately protected by approved methods to maintain the rating of the building element penetrated. This is especially true in concealed spaces where fires can start and grow unchecked. Penetrations into and through fire-rated partitions, floors, and other components are subject to the limitations established in IRC Section R302.4, Dwelling Unit Rated Penetrations.

New electrical installations and alterations to existing electrical systems must be made under a permit issued by the jurisdiction having authority. All electrical components, materials, and equipment must be approved and must be listed and labeled for the specific application in which they are used (see Figure 16-2). All such components, materials, and equipment must bear the label of an approved agency and, like mechanical and plumbing equipment, must be installed and used in strict accordance with the manufacturer's listing and installation criteria as directed by IRC Section E3403.3.

The IRC establishes general requirements for equipment. Electrical equipment intended to interrupt current at fault levels must be rated for a minimum of 10,000 amperes. The IRC requires equipment labeled "for dry locations" or "for indoor use only" to be adequately protected during building construction. Generally, the integrity of electrical equipment must be maintained. This means that the internal parts of all electrical equipment such as busbars, wiring terminals, and

Figure 16-2

A listed, plastic electrical outlet box for a bathroom receptacle.

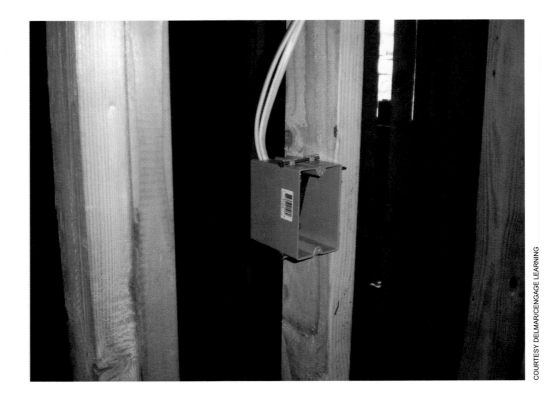

insulators must be adequately protected from contaminants such as paint, plaster, and other debris. Electrical equipment must be protected from physical damage as well. For safety reasons, energized parts of electrical equipment and systems operating at greater than 50 volts must be adequately protected against accidental contact by approved enclosures to prevent serious injury or death.

Equipment must be installed in such a way that it can be serviced for repair and maintained. IRC Section E3405.2 establishes working clearances in front of energized equipment and panels of 36 inches (914 mm) in depth (measured in the direction of access) by 30 inches (762 mm) in width. The IRC also establishes a minimum vertical clearance of 6.5 feet (1,981 mm) above the floor or service platform. Additionally, IRC Section E3405.3 requires a dedicated space above the panelboard, extending to 6 feet (1,829 mm) above the height of the panel or to the ceiling. The dedicated space must be at least as wide as the panel. Generally, no other equipment (e.g., mechanical or plumbing systems) may occupy this space. Access to the working space must be maintained.

IRC Chapter 34 also establishes minimum requirements for electrical conductors and connections. IRC Section E3406 provides basic requirements for all such components and identifies both required wire types and gauges (in American Wire Gauge or AWG sizes). The IRC addresses terminals, splices, continuity, and other important safety criteria. Conductors must be properly identified per Section E3407 to ensure that they can be easily recognized when accessed by qualified installers and inspectors. Generally, the wire size is identified by a color-coded marking scheme in which the outer insulation around the wire varies by color and by a series of stripes that encircle the wire and run along its length. Such schemes enable electricians and inspectors to identify the size, type, and function of the wire by visual inspection.

Services: IRC Chapter 36

This IRC chapter provides code criteria for electrical services, defined as the conductors and equipment used to deliver energy from the electrical utility provider to the building wiring. Electrical energy is generally delivered to a dwelling in one of two ways: by overhead services or underground. As expected, the IRC provides criteria for both. One- and two-family dwellings may have only one service per the requirements of IRC Section E3601.2 and an electrical service may not be provided through another dwelling.

Figure 16-3

Table E3602.2 Minimum service load calculation.

TABLE E3602.2
MINIMUM SERVICE LOAD CALCULATION

LOADS AND PROCEDURE
3 volt-amperes per square foot of floor area for general lighting and general use receptacle outlets.
Plus
1,500 volt-amperes multiplied by total number of 20-ampere-rated small appliance and laundry circuits.
Plus
The nameplate volt-ampere rating of all fastened-in-place, permanently connected or dedicated circuit-supplied appliances such as ranges, ovens, cooking units, clothes dryers not connected to the laundry branch circuit and water heaters.
Apply the following demand factors to the above subtotal:
The minimum subtotal for the loads above shall be 100 percent of the first 10,000 volt-amperes of the sum of the above loads plus 40 percent of any portion of the sum that is in excess of 10,000 volt-amperes.
Plus the largest of the following:
One-hundred percent of the nameplate rating(s) of the air-conditioning and cooling equipment.
One-hundred percent of the nameplate rating(s) of the heat pump where a heat pump is used without any supplemental electric heating.
One-hundred percent of the nameplate rating of the electric thermal storage and other heating systems where the usual load is expected to be continuous at the full nameplate value. Systems qualifying under this selection shall not be figured under any other category in this table.
One-hundred percent of nameplate rating of the heat pump compressor and sixty-five percent of the supplemental electric heating load for central electric space-heating systems. If the heat pump compressor is prevented from operating at the same time as the supplementary heat, the compressor load does not need to be added to the supplementary heat load for the total central electric space-heating load.
Sixty-five percent of nameplate rating(s) of electric space-heating units if less than four separately controlled units.
Forty percent of nameplate rating(s) of electric space-heating units of four or more separately controlled units.
The minimum total load in amperes shall be the volt-ampere sum calculated above divided by 240 volts.

Of paramount importance is the ability to disconnect the entire electrical system installed within the dwelling. The provisions of IRC Section E3601.6.1 require the disconnect to be plainly and permanently marked and it must be readily accessible. For the purposes of the electrical code, "readily accessible" means that the disconnect and other service equipment must be accessed quickly and without the necessity to obtain a ladder or other special equipment. In the event of a fire or injury, the speed at which the electrical service can be accessed and disconnected can mean the difference between life and death.

Electrical services must be sized to adequately serve the dwelling in which they are installed. The IRC requires ungrounded service conductors to be no smaller than three-wire, 100 amperes for one-family dwellings. Other installations, such as in residential accessory structures, must be at least 60 amperes. The service load calculations are determined in accordance with IRC Table E3602.2, as seen in Figure 16-3. Ungrounded service conductors serving less than 100% of the dwelling unit load must be computed as required for feeders per the requirements of Chapter 36, Branch Circuit and Feeder Requirements.

Overhead service drops and service conductors must conform to the provisions of IRC Section E3604 and to the requirements of the local utility provider. Open conductors and multiconductor cables without an overall outer jacket must be installed with appropriate clearances from openable windows, doors, and decks. Service conductors installed over a roof must be at least 18 inches (457 mm) and as much as 8 feet (2,438 mm) above the roof, depending upon the pitch of the roof and the location. These clearances are an important safety measure designed to prevent someone from inadvertently coming into contact with the overhead conductor which could cause serious injury or death.

Service drop cables that are supported on and cabled together with a grounded bare messenger wire must have a minimum clearance above grade of 10 feet (3,048 mm) at the electrical service entrance and the lowest point of the drip loop. Overhead conductors located above residential property and driveways must have a vertical clearance of 12 feet (3,658 mm), and 18 feet (5,486 mm) vertical clearance is required above public streets, alleys, roads, and parking areas subject to truck traffic.

Services must be adequately grounded to ensure safety to humans. In fact, it is safe to say that the proper grounding of an electrical service is one of the most important safety aspects of the electrical system. The ground is an intentional connection between the electrical wiring and service to the earth. This ground provides a path of least resistance should an electrical malfunction occur. The ground is made with a grounding electrode conductor that is connected to a grounding electrode system approved under the code. The ground conductor must be carefully sized to ensure it will function as intended under a load.

Figure 16-4

A rebar stub protruding out of the foundation wall to be used for connecting the service ground. Note the orange sticker indicating the concrete-encased grounding electrode was inspected and approved.

Grounding electrode systems are defined in IRC Section E3608 and the IRC establishes specific code criteria for each type of the system chosen. Grounding electrode systems can be provided in a number of ways. Where there exists a metal underground water pipe at least 10 feet (3,048 mm) in length, the ground may be bonded to it to serve as the grounding system. Concrete-encased electrodes (commonly called a **Ufer** ground) may also be used (see Figure 16-4). In this case, the ground is provided by a wire embedded in concrete or by 0.5 inch (13 mm) minimum diameter reinforcing steel bars placed in the footings and foundations that are in contact with the earth. Wires and rebar used in these applications must be at least 20 feet (6,096 mm) in length. Other options exist as well, such as ground rings, rod and pipe electrodes, and plate electrodes.

Bonding of the electrical service and other metallic components of the system is required to ensure electrical continuity and to safely handle fault currents that might occur within the system. Bonding is the intentional connection of all noncurrent carrying metal components in a room or building to raise them to the same electrical potential. This practice ensures that the occupants of a room or building cannot touch two separate metallic items at significantly different electrical potentials to prevent shock, even where the connection to ground is lost. IRC Section E3609 provides the code criteria to safely accomplish the bonding of metal system components. Bonding is accomplished through the use of threaded connections, threadless couplings and connectors, wires attached to bond screws, or by other means. Bonding of all metal pipe installed within a structure (gas and water, for example) is required.

Branch Circuit and Feeder Requirements: IRC Chapter 37

Branch circuits are the conductors that supply electrical energy to outlets, lights, appliances, and equipment. In the case of convenience outlets and general lighting, it is common for multiple outlets and lights to be served on an individual branch circuit. These branches are typically zoned and arranged conveniently according to the general layout of the dwelling. For example, the lighting and outlets for a master bedroom would likely be on the same branch circuit, up to the maximum size (load carrying capacity) of the circuit.

Some branch circuits are specifically dedicated to a single appliance. This occurs where the electrical load requirements for that appliance are fairly high (such as with a clothes dryer or cooking range) or where there is a specific need to isolate the appliance for service and safety reasons. Branch circuits are sized to ensure that they will safely handle the loading requirements for the appliances and equipment they serve. The circuit size for general lighting and convenience outlets in dwellings is 15 or 20 amperes. Some circuits are sized larger (e.g., 30 amperes) to accommodate the increased electrical demands of water heaters, heat pumps, and clothes dryers.

Wire size must be carefully determined to ensure that it will safely carry the current necessary to serve the appliances. The greater the load, the larger the wire must be to accommodate it. IRC Table E3702.13 establishes minimum wire sizes for branch circuits and varies according to the circuit rating (see Figure 16-5). It stands to reason that high-demand equipment and appliances require not only larger circuits but also larger wire sizes to safely deliver the electrical energy necessary to operate it without the wiring becoming overloaded, a significant fire hazard. The minimum quantity, size, and distribution of branch circuits are determined by the overall demand of the electrical system.

Feeder conductors are also addressed in Chapter 37. Feeder conductors that do not serve 100% of the dwelling unit load within the dwelling must comply with the requirements of IRC Section 3704. Like circuits and other components, feeders must be properly sized and installed in such a way that they

Figure 16-5

Table E3702.13 Branch circuit requirements—summary.

TABLE E3702.13
BRANCH-CIRCUIT REQUIREMENTS—SUMMARY[a,b]

	CIRCUIT RATING		
	15 amp	20 amp	30 amp
Conductors: Minimum size (AWG) circuit conductors	14	12	10
Maximum overcurrent-protection device rating Ampere rating	15	20	30
Outlet devices: Lampholders permitted Receptacle rating (amperes)	Any type 15 maximum	Any type 15 or 20	N/A 30
Maximum load (amperes)	15	20	30

a. These gages are for copper conductors.
b. N/A means not allowed.

will not become overloaded. Overcurrent protection for feeders is required by an approved device installed at the point where the feeder conductor first receives its supply, per IRC Section E3705.5. Overcurrent devices must be readily accessible, must be located to avoid physical damage, and must not be located in prohibited locations such as bathrooms or clothes closets.

Power and Lighting Distribution: IRC Chapter 39

Chapter 39 deals with the distribution of lighting and receptacles throughout the dwelling. The requirements of the IRC are intended to cause all required areas of the dwelling to have adequate lighting and receptacles located conveniently throughout the structure. Receptacle placement is based on the idea that a lamp or appliance can be plugged in anywhere along the wall of a room without the need for an extension cord. Improper use of extension cords is a major cause of electric shock and fires in dwellings.

Spacing and placement of outlets and receptacles is regulated under the provisions of IRC Section E3901.2 to ensure that such items are both convenient and functional. Family rooms, dining rooms, living rooms, parlors, and related habitable areas are required to have receptacles located so that no point along the floor is more than 6 feet (1,829 mm) from an outlet. This translates to a maximum distance of 12 feet (3,657 mm) from outlet to outlet. Wall spaces 2 feet (610 mm) or greater in width, fixed panels in exterior walls, and fixed room dividers and counters are included in this measurement.

Receptacles used for small appliances in kitchens, pantries, and similar areas serve as the required wall outlets described previously, and are on 20 ampere circuits. Small appliance receptacles may serve no other outlets. Countertop receptacles must be spaced generally at intervals not greater than 4 feet (1,219 mm), and within 2 feet (610 mm) of the end of a counter. Kitchen islands and peninsulas must also be provided with receptacles under the provisions of the code (see Figure 16-6). Additionally, laundry areas, hallways, bathrooms, basements, garages, and other spaces must be provided with at least one receptacle.

Kitchens, bathrooms, garages, accessory buildings, and outdoor receptacles (generally, any that are located in close proximity to water or earth) must be provided with ground-fault circuit interrupter (GFCI) protection. This important safety feature provides instantaneous interruption of the flow of electricity where a ground fault occurs. The GFCI opens the circuit when current flow in the "hot" (ungrounded) wire is 5 milliamperes greater than the return current in the "neutral" (grounded) wire. This small difference in current flow indicates that the appliance or tool plugged into the GFCI-protected circuit is defective. While the user may experience a mild tingling sensation, circuit interruption protects the user from a serious shock hazard. Bedrooms must be provided with combination arc-fault interrupters. Such devices detect and open the circuit when damaged wiring has series arcing (within a single conductor) or parallel arcing (between two conductors).

Lighting outlets are also regulated under the electrical provisions of IRC Section E3903. All habitable rooms and bathrooms are required to have wall switch-controlled lighting outlets, except that rooms other than kitchens and bathrooms

Figure 16-6

A convenience outlet mounted on a kitchen peninsula.

COURTESY DELMAR/CENGAGE LEARNING

Figure 16-7

The business end of a residential electrical panelboard.

COURTESY DELMAR/CENGAGE LEARNING

may have a switched outlet that would operate a lamp or similar plug-in device. The IRC also allows occupancy sensors to be used to control lighting within rooms under two conditions: They may be used in addition to the required conventional wall switch or they may be used as a stand-alone switch if located in the customary switch location and as long it is provided with a manual override switch.

Additionally, electrical panels and cabinets are regulated under the code provisions of Chapter 39. The location of the panels containing circuit breakers is regulated and is subject to the clearance requirements described in the previous sections. Electrical panelboards serve as the heart of the electrical system and all circuits for lighting, convenience outlets, equipment, and appliances are routed from the panel to their final location (see Figure 16-7).

Energized panels supply and distribute electrical power sufficient to severely shock or even electrocute human beings. As such, the IRC imposes strict rules that must be closely followed to ensure a safe installation. Electrical panels must be effectively closed to prevent objects from being placed inside. Unused openings into the panel must be covered. Cables that enter the panel must be secured and installed in such a way that they will not be subjected to damage from abrasion or other movement. Panels must also be covered when energized to prevent accidental contact.

Devices and Luminaires: IRC Chapter 40

This IRC chapter regulates the devices typically used in the day-to-day activities within the dwelling. Light switches, light fixtures, receptacles (convenience outlets), and other items all find their place here. General use switches, such as those used to operate lighting and to power equipment, are required to be rated for the intended application. General use and motor circuit switches must be clearly marked so that the "on" and "off" positions are easily identified. Single use switches (those that control motors and other equipment) must be oriented so that the "up" position of the switch handle is the "on" position.

Metal enclosures must be grounded and where plastic or other nonmetallic enclosures are used with metal **raceways** (conduit), grounding continuity must be maintained. Metal boxes must be effectively bonded per Section E3609.4; and where wiring is placed in nonmetallic boxes, wiring must be accomplished with a wiring method that utilizes a grounding conductor. Switches must be located so that they have ready access for operation and must not be located more than 6 feet 7 inches (2,007 mm) above the floor or working platform.

Electrical receptacles, or plugs, must be rated and approved for the intended use as are switches and other components. General purpose receptacles must be rated for 15 or 20 amperes, 125 volts. Special purpose receptacles (for room air conditioners, dryers, ranges, and such) may be rated from 15 amperes, 250 volts to 50 amperes, 250 volts. Where two or more receptacles are placed together, they must conform to the requirements of IRC Table E4002.1.2 (see Figure 16-8). Receptacles must be of a grounding type. Receptacles intended for use in damp locations must meet the requirements of IRC Section E4002.8. Enclosures for 15 or 20 ampere receptacles installed in damp locations must be listed as weather resistant. Receptacles are prohibited in or above showers and bathtubs for obvious reasons.

Lighting fixtures, or luminaires as they are referred to in the IRC and the NEC, must be installed and located in a safe manner. Many lighting fixtures in use today have high-intensity bulbs that become quite hot during operation. Thus, luminaires must be located in such a way that combustible materials will not heat to greater than 194° F (90° C). Lighting fixtures must be constructed in such a way that there are no exposed conductive parts once energized. As with all other electrical components, luminaires must be grounded.

Where recessed incandescent lighting is used in an insulated space, it must be provided with thermal protection and listed as thermally protected per IRC Section E4003.5.

Figure 16-8

Table E4002.1.2 Receptacle ratings for various size multi-outlet circuits.

TABLE E4002.1.2
RECEPTACLE RATINGS FOR VARIOUS SIZE
MULTI-OUTLET CIRCUITS

CIRCUIT RATING (amperes)	RECEPTACLE RATING (amperes)
15	15
20	15 or 20
30	30
40	40 or 50
50	50

Where luminaires (light fixtures) will be installed in wet locations, they must be specifically listed for such use and must be labeled accordingly. Cord-connected or pendant light fixtures are prohibited in or above shower and bathtub areas. Fixtures located in these areas must be approved for wet locations. Where light fixtures are placed in clothes closets, they must be of a completely enclosed type and must be located at least 12 inches (305 mm) (measured horizontally) from the nearest edge of the space above a shelf. Such fixtures can be either surface mounted or recessed and must be of an approved type per Section E4003.12.

Certain types of lighting fixtures, such as track lighting, have special requirements associated with them to ensure a safe operation. Track lighting is prohibited in certain areas such as wet or damp locations and where they are concealed or subject to physical damage. Tracks used for such lighting must be securely mounted and properly grounded.

GREEN OPTIONS

Energy consumption in the United States is staggering. In 2007 alone, 40.2 trillion British thermal units (BTU) of energy were consumed from all sources to generate the electricity needed for residential, commercial, and industrial uses. According to the U.S. Energy Information Administration, renewable energy consumed in 2007 "decreased by about 1 percent between 2006 and 2007, contributing 7 percent of the nation's total energy demand, and 8.4 percent of the United States electricity generation" for the same year.

Because of this enormous demand, any effort to manage energy use and reduce the demand for electrical power is a step in the right direction. Equipment, appliance, and lighting fixture efficiency can contribute significantly to savings for the owner of a green dwelling, and a number of options exist currently. Automated switching and system controls are one important area where significant savings can be realized. According to the U.S. Environmental Protection Agency (EPA), the average household spends $2,200 per year on energy bills, approximately half of which goes toward heating and cooling. Homeowners can save approximately $180 per year by properly setting a programmable thermostat, one example of the automated controls now available for use in dwellings.

Lighting controls also offer additional possibilities. These controls, like occupancy sensors, have long been used in commercial buildings and are now required in most commercial codes. Lighting controls can reduce lighting energy consumption in existing commercial buildings by as much as 50% and by as much as 35% in new commercial construction. This simple technology can also be used in residential construction to reduce lighting energy consumption. In these ways, improved electrical system performance can be easily accomplished and help the green dwelling to do its part for sustainability.

New options also exist in the generation of electricity for residential use. Power generated by wind, hydroelectric means, solar (photovoltaics), and fuel cell technology open up new possibilities for dwelling owners (see Figures 16-9 and 16-10). Many of these possibilities will enable homeowners to reach their goals of owning a net-zero dwelling in the near future.

Figure 16-9

A free-standing, solar photovoltaic panel used to collect the sun's energy for the manufacture of electricity .

COURTESY OF RUSS HOLLAND

Figure 16-10

A DC inverter used in conjunction with a photovoltaic panel to convert solar energy into electricity that can be used within a dwelling.

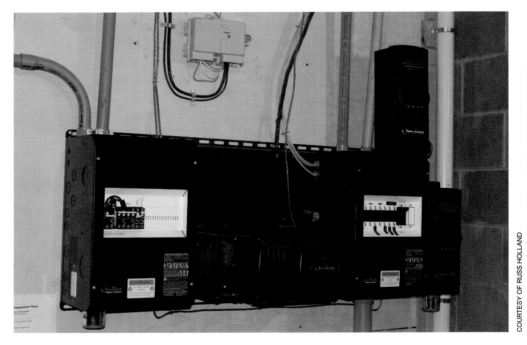

COURTESY OF RUSS HOLLAND

IMPROVING ELECTRICAL SYSTEM PERFORMANCE

As mentioned earlier, electrical energy use and management is beyond the scope of the IRC electrical provisions. Even so, electrical systems can be improved in a number of ways. One of the simplest and most effective ways to improve system performance and increase efficiency is to use less energy. Where the green dwelling is supplied by an electrical utility as most are, the same benefits afforded those who conserve water apply here as well: Use less, then pay less. The less energy needed to operate circuits, lighting, and electrically operated equipment, the better for the environment and one's pocketbook. Electrical performance and efficiency can also be improved by installing systems that *generate* electrical energy. The following sections will explore a few of these options.

Energy-Conserving Fixtures and Appliances

Often, the art of energy conservation and reduced electrical consumption comes down to the simple choice of picking the item that uses the least amount of energy. Motors for blowers, fans, and pumps all consume electricity, yet energy efficiency can vary widely not only across brands but from model to model within the same brand. Simply checking the efficiency ratings for such items is all it takes to save energy, yet how many of us actually take the time to consider total energy used by equipment and appliances before purchasing them?

Many household appliances are labeled with standardized annual energy consumption ratings that make it easy to compare total energy use. Energy Guide labels, as they are known, assist consumers in making informed choices regarding energy efficiency when purchasing appliances. Not all equipment is marked in the same way; thus, some effort and education is needed to make comparisons.

Some energy-saving items have already been discussed in Chapter 13 of this text under the topic of energy conservation. Energy-efficient lighting is one of the simplest ways to reduce energy use within the dwelling. Selecting household appliances and HVAC equipment with the EPA Energy Star label is another easy way to select highly efficient appliances and use less (then pay less). Following are a few more ways to reduce energy consumption within the green dwelling.

Green Decisions and Limitations

Shall I incorporate energy-conserving appliances, equipment, and fixtures into the design of my green dwelling?

Are there restrictions within the IRC that affect these decisions?

Options

When selecting energy-conserving electrical equipment and appliances that will meet the requirements of IRC Part VIII, consider the following additional green options.

Occupancy sensors: Lighting can be controlled automatically through the use of occupancy sensors. Occupancy sensors provide automatic control of lighting through the use of

Why It's Green

Equipment, appliances, and fixtures that use less electricity directly reduce the demands placed on the electric utility grids in various regions around the United States. Such reductions are regarded as sustainable because they reduce the demand for the energy needed to produce the electricity, much of which still comes from fossil fuels or coal. Lower energy consumption also means lower operating costs for the owner of the dwelling, which is always a welcome addition.

Figure 16-11

An occupancy sensor conveniently and attractively housed in a wall switch. Such devices sense when people are present in a room and turn lights off and on as needed.

COURTESY DELMAR/CENGAGE LEARNING

a sensor that detects when a person is present in a room (see Figure 16-11). Once the person leaves the space, the occupancy sensor shuts off the light after a predetermined period of time. These devices have been used in commercial settings for years and are now required in commercial occupancies by many energy conservation codes. Because lighting is used only when necessary and cannot be left on accidentally when in automatic mode, significant energy savings can be realized. Occupancy sensors can be sensitivity adjusted so that they are not triggered inadvertently.

Daylight and photocell detectors: Some detectors can control lighting through the use of photocell and other daylight detection devices. Such devices can automatically shut off or dim lights when daylight reaches certain threshold levels. Energy can be conserved in this way because no more lighting is used than is necessary based on conditions. In winter months or during extremely cloudy conditions, such detectors allow full lighting. These types of detectors are typically provided with a manual override which allows lights to be switched on regardless of conditions so that the user still has control of the lighting system if needed.

Centralized computer controls: Let's face it: Computers are everywhere. They are found in our cars, homes, and even our pockets and purses with the advent of the PDA and other similar devices. Computer technology has enabled improvements in all areas of our lives. Energy consumption can now be managed through the use of a computer as well. New technology makes it possible to control systems such as HVAC, lighting, water heating, and solar power through the use of a computerized energy management system.

Such systems use software to control how and when energy-consuming appliances operate. Equipment can be programmed to shut off or operate less frequently during times when the dwelling is not occupied (such as when the kids are at school and the parents are at work) and can also be programmed to operate in a similar way when the occupants of the dwelling are asleep. Water heating can occur at times when energy is less costly and some programs can even display how much energy is used by specific appliances and systems, offering usage tips that allow for increased energy efficiency through a comprehensive energy analysis. This exciting technology is poised to become standard equipment in green dwellings of the future.

Electrical Energy Generating Equipment

Previously, we have discussed solar photovoltaic and wind turbine systems for the exciting options they offer and for their contribution toward sustainability. These systems actually generate electricity through the free power provided by the sun and wind and allow electrical energy to be stored in batteries or placed back into the power grid. This results in a potential energy *credit* for the green dwelling owner instead of a bill (in effect, the utility *pays you* instead of the other way around).

Other power generating systems, however, warrant further consideration. Such systems utilize **hydropower** (power generated by moving water), fuel cell technology, and other means to accomplish the same result. The last few sections explore some of this new technology as well as the code considerations imposed by the IRC.

Green Decisions and Limitations

Shall I provide an electrical energy generating system in my green dwelling?

What IRC provisions will impact this decision?

Fuel Cell Option

The U.S. Department of Energy describes fuel cell technology as "an important enabling technology for the hydrogen economy" and that fuel cell technology has the capability to "revolutionize the way in which we power our nation." Such technology utilizes hydrogen as a fuel source to generate electricity, and has the power to greatly reduce our dependence on fossil fuels. The process by which hydrogen is produced is clean and the by-products of its production are oxygen and heat. Hydrogen power also can be used to operate vehicles; and much research is being done to make this a practical possibility.

Today's fuel cell technology makes it possible to obtain electricity through packaged equipment small enough to be mounted in a residential application (see Figure 16-12). Hydrogen generating appliances used to generate electricity and to dispense fuel are allowed under the provisions of the IRC. Thus, such futuristic technology can be incorporated into the green dwelling today!

Why It's Green

Systems and devices that generate electrical energy directly reduce the amount of electricity needed from the electrical utility. This in turn reduces the amount of energy needed by the utility provider to produce the electricity that results in less fossil fuels being consumed to generate the electricity in the first place. Such alternative forms of energy are considered renewable because they are clean and because energy sources such as solar power or hydrogen are essentially limitless and can be easily produced or replaced.

TAKE NOTE!

Hydrogen generating appliances are regulated heavily in multiple areas of the code. Sections M1307.4 and M1904 of the IRC regulate gaseous hydrogen systems, as do the International Fuel Gas Code, the International Fire Code, and the International Building Code. Where such systems will be installed in the green dwelling, be sure to talk with the code official in the jurisdiction to coordinate a code-compliant, safe installation.

Figure 16-12

A residential fuel cell generator.

COURTESY OF CERAMIC FUEL CELLS LIMITED

CHAPTER 16 APPLIED TO THE SAMPLE PROJECT

Chapter 16 pertains to electrical systems used to deliver and distribute energy to and within one- and two-family dwellings. Chapter 16 provides general code criteria for the installation of electrical systems and also addresses power distribution and other components of the electrical system. It provides specific code criteria for the installation of electrical services, circuits and feeders, and luminaires. The following analysis will demonstrate the application of some IRC code provisions for electrical systems with consideration given for the proposed scope of work and our design philosophy.

Site Considerations

The final consideration for the site on which our sample project is located relates to the electrical service. Let us assume for this project that the local electrical utility provides an underground electrical service as opposed to an overhead service. Recall one of the design philosophies for the project was to limit the disruption to the site to the maximum extent possible. Thus, to the extent the IRC allows, we can combine the wiring for the electrical service in the same trench as the water and sewer service. We will check with the local building department for both the minimum depth and separation requirements as some of these vary based on geographic conditions. This will limit the excavation necessary to place the utilities for the project.

Building Considerations

Since our foundation is likely to require reinforcing steel to accommodate the structural loading requirements of the rammed-earth walls, a concrete-encased electrode makes sense for the grounding electrode, and can easily be installed per the provisions of Section E3608.1.2. The service size, wiring types and gauges, receptacle locations, and luminaires will all be installed per the requirements outlined in the IRC.

The scope of work for the project specified that high-efficiency lighting would be used wherever possible throughout the project. Recall that the energy conservation requirements of IRC Chapter 11 (Chapter 13 in *Going Green with the IRC*) requires at least 50% of the permanently installed lighting fixtures to be provided with high-efficacy lamps (bulbs), a requirement we will easily meet. High-efficacy lamps can be tube type or compact fluorescent bulbs, LED, or other lamps meeting the definition provided in the IRC.

Finally, the project specifications outlining the scope of work state that we will incorporate a photovoltaic system into the design. These systems are outside the scope of the IRC electrical provisions but are addressed in detail in the 2008 National Electrical Code. NFPA 690 addresses photovoltaic (PV) systems and provides all necessary considerations to ensure a safe, code-compliant installation. Where the PV system is roof mounted, the additional weight of the panels or tiles must be considered in the structural design of the roof. Where ground mounted, the panels must be located in a manner that will ensure not only that they will function properly but also, like the ground-source heat pump and loop piping, that the installation meets all local regulations for setbacks, height restrictions, and other restrictions.

CHAPTER SUMMARY

Chapter 16 of the IRC addresses the various code requirements for electrical systems installed in dwellings. It provides standards for materials and equipment as well as requirements for construction methods to ensure that electrical appliances, luminaires, and systems are safe and suitable for their intended purpose. Following are highlights of the chapter.

- This chapter addresses requirements for residential electrical systems from Part VIII of the IRC, which includes selected material from IRC Chapters 34 through 43.
- The material presented in Chapters 34 through 43 is based on the 2008 edition of the National Electrical Code.
- The material presented here is unique in that it is reprinted under license from the National Fire Protection Association (NFPA), rather than having been developed by the International Code Council.
- Some chapters are not addressed because they are not relevant to the discussion of green dwellings or because they offer no practical green options.
- Electrical systems supply electrical energy to the dwelling that is used to operate appliances, equipment, lighting, and outlets.
- The electrical code provisions address general electrical provisions, services, branch circuits and feeders, power and lighting distribution, luminaires, and more.
- The efficiency and performance of traditional electrical systems can be improved by selecting electricity-conserving fixtures and appliances (commonly called high-efficiency appliances).
- Solar photovoltaic, wind power, fuel cells, and other energy generating systems make excellent choices for green dwellings and are regulated under the provisions of the IRC and the National Electrical Code.
- The manufacturer's installation guidelines for all electrical fixtures, appliances, equipment, and components must be carefully observed.

It is my sincere hope that *Going Green with the IRC* was an enjoyable text and will continue to be useful in your quest to incorporate green construction into every project. As we have seen, the building code is, out of necessity, a comprehensive and complex document that often takes many years for a code professional to fully understand.

There is no expectation that all of the subjects covered in this text be understood after one reading. Refer back to its various sections as often as needed during the planning, permitting, and construction processes to reduce risk of errors. Ask questions. Use the text as a resource at each stage of development. Above all else, establish a relationship with the code official in your area. This person can be an invaluable resource to you when things go wrong (and things will invariably go wrong), so it makes good sense to capitalize on a well-founded relationship.

The Changing Face of Building Codes

An important and exciting change is afoot in the world of building codes. While the code has not historically addressed matters such as how sustainable a construction practice is or, say, the minimum required level of "green" for a particular material to be deemed suitable for use, it is widely recognized that these issues are a growing concern for much of society. Those who develop and enforce the codes are increasingly aware of this fact and are making efforts to incorporate green thinking into the building code and its companion publications.

Just as the code has adapted over time to address health and safety matters that have come to the forefront, such as the elimination of lead from water supplies and the addition of code provisions to address radon, so it would seem the code is positioned to do the same in the case of green building. The code does not eliminate older code provisions to make way for the new; rather, it adapts and adds requirements as necessary based on new thinking and the identification of societal needs that were not necessarily recognized or even known in years past.

Society is now focused on a broader and a more comprehensive set of issues related to development and construction, many of which transcend the traditional structural, health, and life safety concerns that have been the main focus of the code for decades. In short, where the code has always been concerned with the impact of the environment on buildings and other structures, we are now increasingly focused on just the opposite—the impact that *buildings* have on the environment. It is this important shift in thinking that fuels the move toward sustainable development and construction practices.

There is now an increased awareness and a deeper sense of responsibility toward the impacts that construction has on the environment, from the harvesting practices used to collect timber to the manufacturing processes used to make construction materials. While it is yet to be seen whether matters related to green building will one day end up as standard provisions in the IRC, one thing is certain: This new emphasis on sustainability and green building will certainly shape and influence the discussion. It seems likely, then, that many of the issues we are discussing today as green building methods may one day soon be actual concerns of the IRC. It is my hope that such matters will one day be as widely accepted and as routinely enforced as matters of life safety and structural soundness. Only time will tell. . . .

The Author's Perspective

Until now, I have purposely avoided expressing my personal preferences on the subject of green building because I did not want to sway the reader's opinion toward one construction method over another, nor did I want to imply that one green material was preferred or somehow superior to another material. As I have alluded to in the text, just exactly what the term "green building" means is still the subject of much debate. There is, at this time, no universally accepted definition of what "green" or "sustainable" means and, thus, there is no absolute right or wrong. It would be presumptuous and perhaps even irresponsible of me to suggest that I somehow have all the answers on this topic. As I stated in the Preface, I make no attempt to settle the debate on this matter.

Still, I would be remiss if I did not at least attempt to provide some guidance here, as there are a number of principles on green building and sustainability that have risen to the surface and have gained fairly widespread acceptance to date. The following not only reflect those notions of sustainability that seem to have become increasingly accepted as a necessary part of the discussion but also reflect my opinion on the matter, because they make good sense.

- **Reduce waste where possible.** Construction produces an extraordinary amount of waste. If you have ever driven by a residential construction site and noticed the large piles of debris stacked in the front yard or the massive dumpster filled to the brim with waste materials, you know what I mean. Any efforts made to reduce the amount of construction waste headed for landfills are worthwhile. Estimate material needs carefully and purchase no more materials than are absolutely necessary for the job. Measure twice, cut once to reduce errors that contribute to the waste stream. Recycle building materials where possible and donate leftover materials in good condition to charitable organizations where these options exist in your area rather than throwing them away.
- **Utilize salvaged materials where possible.** Salvaged products not only reduce waste but also reduce demands for the resources needed to manufacture new materials and equipment. A number of cities now have architectural salvage stores and renovation centers from which one can purchase treasures such as antique doors, fireplace surrounds, and decorative wood trim. I love nothing more than to see these types of products given new life by incorporating

them into a modern construction or renovation project. Utilizing these treasures is an important way to preserve a bit of our heritage and soften the blow to the environment by using wood products that have already been taken from nature.

- **Utilize materials made from the by-products of manufacturing processes, recycled materials, or other waste products where possible.** Products made from manufacturing by-products and other waste products maximize the use of resources and capitalize on the embodied energy present in the waste material. Remember that considerable energy was likely expended in the manufacture of the source product to begin with, so why not use every bit of the material possible? There are many such products available today, including batt insulation made from denim waste, end-jointed lumber, and roof coverings made from discarded tires. Basically, there are many creative people taking what you and I throw away and forming it into useful construction materials. It makes sense to use these products where available.

- **Use local where possible.** Consider the big picture when choosing a material. Does it make sense to ship an otherwise green product from across the globe when a similar product might be available locally? Local materials are readily available and the impact to the environment from the transportation of the material to the job site is lessened as well. It is also likely that local builders and design professionals are familiar with the material, making it easier to incorporate it into the project. Also remember that sustainability includes more than just matters related to the environment. The economics of green are just as important. Supporting the local economy is an essential part of sustainability that, in my opinion, often gets overlooked.

- **Maximize energy efficiency.** It goes without saying that energy conservation makes sense for everyone. Energy savings benefit the environment and our pocketbooks. Whether building a new dwelling or remodeling an existing one, efforts made to improve the efficiency of the building envelope and fenestrations, building equipment, appliances, and lighting can pay big dividends.

- **Simply do what you can.** No matter how badly we might desire to include every possible green material and method in our construction project, the simple truth of the matter is that nearly all of us are bound by the limitations of a budget. It is a rare set of circumstances where "cost truly is no object" and for those fortunate enough to be in that position, my hat is off to you. For the rest of us, the realities of a budget will mean we have to pick and choose those green elements that are most important to us based on our own needs and beliefs—and that is okay! The tendency for many of us is to think that our contributions will not make a difference, but I assure you that they will. Even where budgets are very limited, small steps can be taken to make a project a little more green. For example, if every household in the United States did something as simple as switch to compact fluorescent or other high-efficiency bulbs where possible, the energy savings would be extraordinary. Again, the answer here is to do what you can reasonably afford to do and be proud of your contribution to a greener world.

Challenge to Code Officials

Finally, I want to offer a challenge to my fellow code officials everywhere. There is no question that certain green construction materials and methods will take us well outside our comfort zones. I sleep best when, in the course of performing my job duties, the construction world around me is familiar and conventional, when I am in that comfortable place where everything is in my control and where I totally understand the methods and materials being used, as with conventional construction. However, green technology, materials, and construction methods are evolving at an extraordinary pace, far faster than the codes can keep up and certainly faster than we can effectively learn about them. Thus, it is up to us to ensure that we do everything in our power to remove obstacles and encourage innovation in this exciting area.

If alternative means and methods are an area where you have limited experience or if they simply make you uncomfortable because they deal with that nebulous "gray area" that seems to be an ever-present part of code administration, I encourage you to interact with and talk to other code professionals. In my experience, there is always someone out there who has dealt with a similar situation and is willing to share lessons learned. Familiarize yourself with the terminology, the materials, and methods currently being used; in short, learn all you can about green building so your knowledge base can keep pace with the rapidly changing industry.

I am in total agreement that no matter how green or sustainable a particular product is, we should never compromise life safety or structural soundness. However, our challenge is to find ways that not only allow the integration of green methods into the project but also do so in a way that preserves or enhances life safety and promotes structural soundness. Now is the time to roll up our collective sleeves and become a part of the solution—a willing and committed partner in the process rather than an obstacle. I invite your correspondence on the matter.

In Closing

Thank you for purchasing *Going Green with the IRC*. It demonstrates your commitment not only to green and sustainable construction practices but also to building safety, which are both important and admirable causes. I wish you the very best in your quest to integrate green building with the International Residential Code. For more information, please visit www.goinggreenwiththeIRC.com, where you will find electronic versions of the forms you see in the text, building code–related tips, plus a variety of helpful information. Good luck!

Yours in green building,
Scott Caufield, CBO

learning objective

To know and understand how the utilization of existing buildings propels the idea of green and sustainable construction forward and to understand how existing buildings are related to the International Residential Code (IRC).

The greenest brick is the one that's already in the wall.
—Steve Mouzon, Architect

The above quote is both simple and poignant. It represents what in my opinion is perhaps the most important opportunity for green building available today—to address, preserve, and utilize the enormous inventory of existing buildings in the United States. Many of these existing dwellings are beautiful reminders of eras long past, when houses were individually crafted by artisans who took pride in their workmanship and whose efforts created streetscapes of enduring beauty like the one seen in Figure A-1.

More important, these words capture the essence of what those of us interested in green building know to be true: that our existing building inventory symbolizes the enormous pool of resources we have already taken from nature. As a result, the amount of embodied energy contained within this inventory is mind boggling, not to mention the good old-fashioned manual labor which has already been put to good use to create the many buildings in which we live, work, and play. It only makes sense to try to preserve as much of this inventory as is practical.

According to the United States Green Building Council (USGBC), there are some tens of millions of existing buildings in the United States. It stands to reason that a good many of these buildings were built prior to the existence of building codes and, certainly, many were built before current energy codes came into existence. Thus, there exists an enormous opportunity to improve what we already have around us.

Figure A-1

This historic streetscape is located near downtown Baltimore, Maryland.

COURTESY DELMAR/CENGAGE LEARNING

The Energy Information Administration, which provides official statistics from the U.S. government, states in the article, "Assumptions to the Annual Energy Outlook 2009," that there are an estimated 72 billion square feet of commercial building space in the United States alone (this includes lodging houses; based on 2003 reported data). This says nothing for the millions of dwellings in the United States, all of which consume energy and require additional resources to maintain and repair. It is easy to see why there is much to be gained if we can effectively manage and improve this inventory.

Fortunately the IRC recognizes this important reality and provides a means to deal with existing buildings and structures in IRC Appendix J. Recall from our discussion in Chapter 15 that a jurisdiction must specifically adopt the provisions of an appendix before they are enforceable as requirements. Where they have not been adopted, they can always be proposed as AMM, assuming this is acceptable to the code official. Let us briefly explore Appendix J.

IRC APPENDIX J: EXISTING BUILDINGS AND STRUCTURES

The opening statement for this appendix chapter is profound, stating "the purpose of these provisions is to encourage the use or reuse of legally existing buildings and structures." It permits work to be performed that is consistent with the purpose of the IRC and also recognizes that work performed under the provisions of Appendix J is deemed to be code compliant. The appendix requires all work performed on existing buildings to be placed in one of four categories: repair, renovation, alteration, or reconstruction. Each successive category introduces a larger scope of work and thus additional requirements to meet the code.

Appendix J recognizes that many existing buildings were built prior to the development and adoption of codes and may not comply with today's code requirements. It also recognizes that it is somewhat impractical to require an existing building, especially an old one, in which a building owner wants to perform work to fully comply with current building codes (see Figure A-2). Appendix J gives code officials a tool to evaluate the scope of work proposed and establishes requirements based on that scope.

Figure A-2

An early 1900s dwelling located in downtown Baltimore, Maryland.

COURTESY DELMAR/CENGAGE LEARNING

Figure A-3

A beautifully detailed entryway and door in a 1900s dwelling.

COURTESY DELMAR/CENGAGE LEARNING

Like the main code language in the IRC, provisions in Appendix J allow the code official to consider alternative materials and authorize the code official to consider alternatives where compliance with the provision of Appendix J is technically infeasible or the costs of compliance are disproportionate to the cost of the proposed work. In no case, however, can the proposed work adversely affect the building, weaken it structurally, or make it unsafe or hazardous.

Other Requirements

IRC Section AJ103.9 requires the code official to meet with the prospective permit applicant when requested to do so, to identify the scope of work and the specific code provisions that apply. Section AJ104 authorizes the code official to require an existing building to be investigated and evaluated by a registered design professional where the proposed scope of work is classified as a reconstruction that includes work to exit systems, life safety systems, and other extensive alterations.

Additional Benefits of Preservation

In addition to the many green benefits afforded through the preservation of existing buildings, there are many aesthetic considerations as well. Existing buildings in some areas contain architectural elements and details that cannot be easily duplicated today (see Figure A-3). In preserving these elements, we allow future generations to appreciate the many buildings that have become a rich part of our heritage while still having a positive impact on the environment.

For a more thorough discussion of existing buildings, see the additional Delmar Cengage Learning title, *Going Green with the IBC,* which will soon be available through www.delmarlearning.com. As an additional resource, the ICC 2009 International Existing Building Code (IEBC) provides comprehensive coverage on this subject and is also available through www.delmarlearning.com.

SAMPLE EVALUATION REPORTS

learning objective

To know and understand the various sections and elements of ICC evaluation reports and how they are related to the International Residential Code (IRC).

CONTENTS OF AN ICC EVALUATION REPORT

An ICC evaluation report (ER, hereafter) is organized into various sections, each of which addresses an important subject area in the consideration of a particular construction material or system for use in a construction project. This data is organized so that code professionals, design professionals, builders, and other interested persons can easily find and extract data relevant to the specification and use of the construction material or system evaluated.

ERs address important subject areas such as product classification, determine which properties or characteristics have been evaluated, and provide the scope of the report itself (see Figure B-1). The following illustration identifies these key ER subject areas and will assist you in becoming familiar with the layout and organization of these important documents.

SAMPLE EVALUATION REPORT

To further our understanding of ERs, consider the sample report in Figure B-2. The report follows the general layout depicted in the previous illustration and is much like any ER the reader may encounter. This sample report applies to a standing seam metal roof panel manufactured by Acme Custom-Bilt Panels. The subject roof panels have been evaluated against the code criteria established in the 2006 editions of both the International Building Code (IBC) and the IRC. They have been identified and approved for use as roof panels in Section R905.10 of the

Figure B-1

An analysis of the various sections of an ICC Evaluation Service evaluation report.

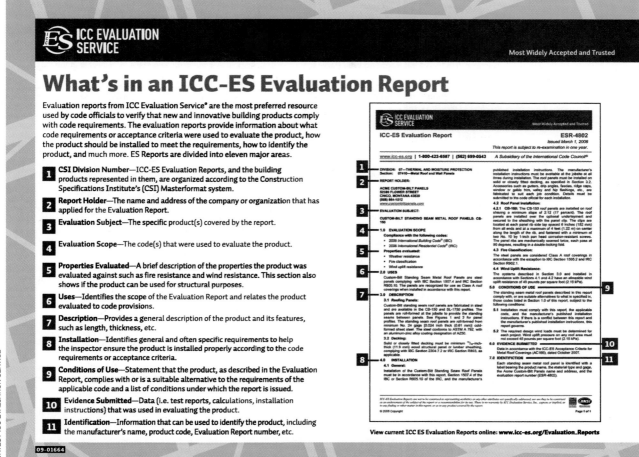

IRC and Section 1507.4 of the IBC. Note that in Section 3.2, the report establishes the decking requirements for use of these roof panels. In this case, solid sheathing or closely spaced sheathing is required as the supporting surface for the panels. The minimum required decking thickness is identified and must comply with the appropriate sections of the IBC and IRC as stated in the report.

Section 4.0 of the report contains important language. Note that this section requires the roof panels to be installed per the conditions of the ER *and* per the requirements of the applicable code. Thus, in order for the roof panels to be regarded as code compliant, they must meet the requirements of both documents. Specific installation criteria are provided in Section 4.2, such as minimum required roof slope and spacing of the required connection clips used to fasten the roof panels into place.

The Conditions of Use outlined in Section 5.0 of the ER are also very important. Note that the ER clearly states that the Acme roof panels comply with and are considered suitable alternatives to the construction methods outlined in the IRC

Figure B-2

A sample ICC Evaluation Service evaluation report. In this case, the report describes the use and outlines the specific conditions of approval for Acme Custom-Bilt standing seam metal roof panels.

ES REPORT™

ESR-4802

Issued March 1, 2008
This report is subject to re-examination in one year.

ICC Evaluation Service, Inc.
www.icc-es.org

Business/Regional Office ▪ 5360 Workman Mill Road, Whittier, California 90601 ▪ (562) 699-0543
Regional Office ▪ 900 Montclair Road, Suite A, Birmingham, Alabama 35213 ▪ (205) 599-9800
Regional Office ▪ 4051 West Flossmoor Road, Country Club Hills, Illinois 60478 ▪ (708) 799-2305

DIVISION: 07—THERMAL AND MOISTURE PROTECTION
Section: 07410—Metal Roof and Wall Panels

REPORT HOLDER:

ACME CUSTOM-BILT PANELS
52380 FLOWER STREET
CHICO, MONTANA 43820
(808) 664-1512
www.custombiltpanels.com

EVALUATION SUBJECT:

CUSTOM-BILT STANDING SEAM METAL ROOF PANELS: CB-150

1.0 EVALUATION SCOPE

Compliance with the following codes:

▪ 2006 *International Building Code®* (IBC)
▪ 2006 *International Residential Code®* (IRC)

Properties evaluated:

▪ Weather resistance
▪ Fire classification
▪ Wind uplift resistance

2.0 USES

Custom-Bilt Standing Seam Metal Roof Panels are steel panels complying with IBC Section 1507.4 and IRC Section R905.10. The panels are recognized for use as Class A roof coverings when installed in accordance with this report.

3.0 DESCRIPTION

3.1 Roofing Panels:

Custom-Bilt standing seam roof panels are fabricated in steel and are available in the CB-150 and SL-1750 profiles. The panels are roll-formed at the jobsite to provide the standing seams between panels. See Figures 1 and 3 for panel profiles.

The standing seam roof panels are roll-formed from minimum No. 24 gage [0.024 inch thick (0.61 mm)] cold-formed sheet steel. The steel conforms to ASTM A 792, with an aluminum-zinc alloy coating designation of AZ50.

3.2 Decking:

Solid or closely fitted decking must be minimum $^{15}/_{32}$-inch-thick (11.9 mm) wood structural panel or lumber sheathing, complying with IBC Section 2304.7.2 or IRC Section R803, as applicable.

4.0 INSTALLATION

4.1 General:

Installation of the Custom-Bilt Standing Seam Roof Panels must be in accordance with this report, Section 1507.4 of the IBC or Section R905.10 of the IRC, and the manufacturer's

published installation instructions. The manufacturer's installation instructions must be available at the jobsite at all times during installation.

The roof panels must be installed on solid or closely fitted decking, as specified in Section 3.2. Accessories such as gutters, drip angles, fascias, ridge caps, window or gable trim, valley and hip flashings, etc., are fabricated to suit each job condition. Details must be submitted to the code official for each installation.

4.2 Roof Panel Installation:

4.2.1 CB-150: The CB-150 roof panels are installed on roofs having a minimum slope of 2:12 (17 percent). The roof panels are installed over the optional underlayment and secured to the sheathing with the panel clip. The clips are located at each panel rib side lap spaced 6 inches (152 mm) from all ends and at a maximum of 4 feet (1.22 m) on center along the length of the rib, and fastened with a minimum of two No. 10 by 1-inch pan head corrosion-resistant screws. The panel ribs are mechanically seamed twice, each pass at 90 degrees, resulting in a double-locking fold.

4.3 Fire Classification:

The steel panels are considered Class A roof coverings in accordance with the exception to IBC Section 1505.2 and IRC Section R902.1.

4.4 Wind Uplift Resistance:

The systems described in Section 3.0 and installed in accordance with Sections 4.1 and 4.2 have an allowable wind uplift resistance of 45 pounds per square foot (2.15 kPa).

5.0 CONDITIONS OF USE

The standing seam metal roof panels described in this report comply with, or are suitable alternatives to what is specified in, those codes listed in Section 1.0 of this report, subject to the following conditions:

5.1 Installation must comply with this report, the applicable code, and the manufacturer's published installation instructions. If there is a conflict between this report and the manufacturer's published installation instructions, this report governs.

5.2 The required design wind loads must be determined for each project. Wind uplift pressure on any roof area must not exceed 45 pounds per square foot (2.15 kPa).

6.0 EVIDENCE SUBMITTED

Data in accordance with the ICC-ES Acceptance Criteria for Metal Roof Coverings (AC166), dated October 2007.

7.0 IDENTIFICATION

Each standing seam metal roof panel is identified with a label bearing the product name, the material type and gage, the Acme Custom-Bilt Panels name and address, and the evaluation report number (ESR-4802).

COURTESY ICC EVALUATION SERVICE

---

and IBC. Also note that the prescribed use requires compliance with the manufacturer's installation instructions; thus, the manufacturer's installation instructions essentially become a part of the code for this application. Given the scope and depth of the topics covered in ERs, it is easy to see why code officials rely so heavily on them as an assurance of code compliance.

SAVE—VERIFICATION OF ATTRIBUTES REPORT

Like the sample report in the previous section, ICC- ES SAVE Verification of Attributes Reports (SAVE reports, hereafter) provide key information to code officials, specifiers, building owners, and other interested persons regarding sustainability claims made by manufacturers of green products (see Figure B-3). SAVE reports provide objective evaluation criteria focused on the materials and processes that go into manufacturing the green product, an increasingly important consideration. SAVE is a voluntary program in which manufacturers seeking to have their

Figure B-3

A sample ICC Evaluation Service SAVE verification of attributes report. In this case, the report verifies the quantity of biobased content in A-1 Insulation.

ICC EVALUATION SERVICE

Most Widely Accepted and Trusted

ICC-ES SAVE Verification of Attributes Report™   VAR-1053

Issued July 1, 2009
This report is subject to re-examination in one year.

www.icc-es.org/save | 1-800-423-6587 | (562) 699-0543   *A Subsidiary of the International Code Council®*

DIVISION 07—THERMAL AND MOISTURE PROTECTION
Section 07 21 16—Building Insulation
Section 07210—Building Insulation

REPORT HOLDER:

A–1 Insulation, Inc.
123A Rocky Road
Asphalt, CA 43210
(123) 765-4321
www.a1insulation.com
jv@a1insulation.com

EVALUATION SUBJECT:

A–1 Insulation

1.0 EVALUATION SCOPE

Compliance with the following evaluation guideline:

ICC-ES Evaluation Guideline for Determination of Biobased Material Content (EG102), dated October 2008.

2.0 USES

A–1 Insulation is a semirigid, low-density, cellular isocyanate foam plastic insulation that is spray-applied as a nonstructural insulating component of floor/ceiling and wall assemblies.

3.0 DESCRIPTION

A–1 Insulation is a two component system with a nominal density of 1.0 pcf (16 kg/m³). The insulation is produced by combining the two components on-site. Water is used as the blowing agent and reacts with the isocyanate, which releases a gas, causing the mixture to expand. The mixture is spray-applied to the surfaces intended to be insulated.

The insulation contains the minimum percentage of biobased content as noted in Table 1.

4.0 CONDITIONS

Evaluation of A–1 Insulation for compliance with the International Codes is outside the scope of this evaluation report. Evidence of compliance must be submitted by the permit applicant to the Authority Having Jurisdiction for approval.

5.0 IDENTIFICATION

The A–1 Insulation spray foam insulation described in this report is identified by a stamp bearing the manufacturer's name and address, the product name, and the VAR number (VAR-1053).

TABLE 1 – BIOBASED MATERIAL CONTENT SUMMARY

% MEAN BIOBASED CONTENT	METHOD OF DETERMINATION
15% (+/- 3%)[1]	ASTM D6866

[1] Based on precision and bias cited in ASTM D 6866.

products evaluated can do so based on established guidelines for sustainability. The SAVE evaluation includes an inspection of the manufacturing facility and processes, along with analysis of testing performed in recognized laboratories.

Note that the SAVE report is laid out in much the same way as an ER. The scope of the report is identified and the specific use for the product evaluated is established as well. Section 4.0, Conditions, contains an important statement that warrants further study. The text states clearly that an evaluation for code compliance is outside the scope of the SAVE report. Thus, the SAVE report can be used only as a measure of how accurate the manufacturer's claims are related to sustainability. In other words, just because a product carries the SAVE distinction does not necessarily mean it will meet the requirements of the IRC, IBC, or other codes. The reader should use an ER for the purpose of determining code compliance.

ALTERNATIVE MATERIALS, DESIGN, AND METHODS (AMM) REQUEST FORM

learning objective

To know and understand the various sections and elements of an Alternative Materials, Design, and Methods Request Form and how they relate to the International Residential Code (IRC).

SUGGESTED USE OF THE AMM FORM

The Alternative Materials and Methods Request Form (AMM form, hereafter) included in this appendix is a useful tool in the preparation and presentation of information, test data, and supplemental materials related to AMM (see Figure C-1). It is organized into various sections, each of which addresses an important subject area in the consideration of an AMM for suitability under the code. These sections are organized in a way that, if the form is filled out thoroughly and completely, will compel the applicant to answer those critical questions that will inevitably be asked by the code official.

Design professionals, builders, and other interested persons can use the form as a guide during the planning and design stages to ensure that they have diligently explored all areas that are concerns of the code with respect to AMM. Remember from our earlier discussion on AMM that a jurisdiction may already have such a form and, of course, you will want to use it if this is the case. Code officials are welcome to use this suggested AMM form as a model on which to develop their own AMM form. In so doing, the jurisdiction can add or emphasize sections that address those areas of particular concern or, likewise, delete those elements that are not useful. The jurisdiction could add its own logo and other identification information or even adapt the form for specific use with green buildings. For more information, visit www.goinggreenwiththeIRC.com, where you will find a downloadable version of this form.

Figure C-1

A sample alternative materials and methods request form. This form may also be used to request modifications of code requirements due to practical difficulties per IRC section R104.10.

**Application for Approval of Alternative to or Modification of the 2009
Edition of the International Residential Code**

Date: _____   Permit No.: _____

Project Name: _____   Project Address: _____

Owner's Name: _____   Phone: _____

Owner's Address: _____

Applicant's Name: _____   Phone: _____

Applicant's Address: _____

Building Department Contact: _____

NOTE TO APPLICANT: Sections R104.10 and R104.11 of the International Residential Code grant the Building Official the ability to consider alternatives to or modifications of the Code in unusual cases and require that individual cases be considered carefully within the context of the requirements of these provisions. Before proceeding with this application it is essential that you read and fully understand the conditions set forth in this application.

This application is specific to and limited to the project identified above.

A.   Section R104.11: Alternate Materials, Design and Methods of Construction:

1.   Pursuant to Section R104.11 of the International Residential Code, the undersigned Applicant hereby requests approval of an alternative to Section _____ of the Code which requires that (cite that portion of the Code from which the Applicant is seeking relief based upon the proposed alternative):

_____
_____
_____
_____

2.   The undersigned Applicant proposes the following alternative to Section _____ of the International Residential Code (provide a detailed description of your proposed alternative):

_____
_____
_____
_____

3.   For the following reasons, Applicant believes that the proposed alternative to Section _____ of the International Residential Code complies with the provisions of the Code and that the material, method or work offered is, for the purpose intended, at least the equivalent of that prescribed in this Code in suitability, strength, effectiveness, fire resistance, durability, safety and sanitation. Include supporting documentation to the extent possible:

_____
_____
_____
_____
_____
_____

Figure C-1 (Cont.) B. Section R104.10: Modification (s):

1. Pursuant to Section R104.10 of the International Residential Code, the undersigned Applicant requests approval of a modification to Section _____ of the Code which requires that (cite that portion of the Code from which the Applicant is seeking relief based on the proposed modification):

_____

_____

_____

_____

_____

_____

_____

2. The undersigned Applicant states the following reason (s) why strict compliance with Section _____ of the Code is impractical or presents extreme difficulty (provide a detailed, specific statement of the reason for your request):

_____

_____

_____

_____

_____

_____

_____

3. For the following reasons, Applicant believes that the proposed modification to Section _____ of the International Residential Code meets the intent of the Code and neither lessens any fire protection requirements of the Code nor compromises the structural integrity of the structure. Include supporting documentation to the extent possible:

_____

_____

_____

_____

_____

_____

_____

_____

_____

_____

Applicant / Owner's Signature and Date:

Figure C-1 (Cont.)

Application Determination

Upon reasonable consideration, the _____ (authority having jurisdiction or AHJ) determines that the above application is:

_____ Denied

_____ Approved Without Conditions

_____ Approved, Subject to the Following Conditions:

_____
_____
_____
_____
_____
_____

Date: _____ Building Official: _____

APPLICANT'S AGREEMENT TO ABIDE BY CONDITIONS

The undersigned expressly acknowledges and agrees that acceptance of this application and any subsequent issuance of a permit (s) based upon the proposed alternative (s) or modification (s), has been made subject to certain conditions which the _____ (AHJ), in its sole discretion, deems necessary. The undersigned agrees to comply strictly with all conditions imposed by the _____ (AHJ). With respect to all permit (s) issued based upon any alternative to or modification of the International Residential Code, the undersigned's failure to comply strictly with all conditions imposed by the Building Department in granting any permit (s) pursuant to this application will render any right to proceed with construction, occupancy or use of any property or premises pursuant to said permit VOID, and will subject the undersigned to immediate revocation of any permit (s) issued in connection with this application. The undersigned and all subsequent owners, occupants or users of these premises claiming any right of occupancy or use of the premises through the undersigned, shall be liable for all costs and expenses, including any reasonable Attorney's Fees and Expert Witness Fees, for enforcement of any condition or term of any permit (s) issued to this application.

The undersigned acknowledges that this agreement does not in any way limit any remedy or right the _____ (AHJ) may otherwise have with respect to enforcement of any of its Codes or Ordinances.

The undersigned acknowledges that any delay by the _____ (AHJ) with respect to enforcing strict compliance with any conditions imposed on any permit (s) issued based upon the proposed alternative (s) or modification (s) shall not be deemed to be a waiver and shall not stop or bar the _____ (AHJ) from enforcing compliance with any conditions, including the _____ (AHJ) right to issue, cease and desist orders and / or to seek immediate relief, as appropriate.

Agreed and Accepted:

Owner's Signature: _____ Date: _____

If Applicant is not the Owner or the Owner's Architect or Engineer:

Applicant Signature / Title: _____ Date: _____

Section A; R104.11, Alternative Means, Design, and Methods of Construction

Section A (1) of the AMM form requires the applicant to identify the particular code section or sections for which the request for the use of alternate materials and methods is made. Section A (2) of the AMM form requires the applicant to specifically outline the proposed alternate method. This section should be thoroughly completed so the code official will understand the nature and scope of the proposed work.

Section A (3) of the AMM form is perhaps the most important section of the form. In it, the applicant must demonstrate how the proposed AMM not only meets the intent of the code but also how it is equivalent to the requirements of the code. Test data, engineering calculations, and ICC ERs or Legacy Product Approval Reports should be attached and any supplemental materials relevant to the proposal should be included. The burden of proof is on the applicant as we have discussed previously. For a review of the minimum items that should be included in the AMM proposal, see Chapter 4 of this text.

Section B; R104.10 Modifications

Another reason this form is so useful is that it addresses requests for modifications that may be pursued under the provisions of IRC Section R104.10. This section provides the building official with authority to grant modifications to the code in individual cases where, in his opinion, there are "practical difficulties" in meeting code requirements and where compliance with a specific code provision is otherwise impractical. Although this important code section has not been the primary focus of this text, it is worth noting here because the form lends itself well to the task of making such requests.

Section B (1) of the AMM form requires the applicant to identify that particular code section for which the request for modification is made. Additionally, Sections B (2) and (3) of the AMM form require the applicant to cite the reason(s) why strict compliance with the code is impractical. More important, the applicant must demonstrate how the modification not only meets the intent of the code but also how it will not reduce or lessen the fire protection or structural requirements of the code. As with AMM, the burden of proof is on the applicant.

Approval and Disclaimer

The AMM form provides an area where the code official can approve, deny, or establish conditions under which an AMM or modification might be approved. This section is important because he can establish specific conditions under which he will allow the modification, say with the addition of an automatic fire sprinkler system, for example. This section captures for the public record important information as to the basis of the approval should some question arise as to why the deviation from the specific code requirement was allowed.

Finally, the AMM form provides a suggestion for the terms of agreement by which the applicant is obligated to abide and an important disclaimer for the jurisdiction. I highly advise all parties utilizing this form to seek the advice of legal counsel to ensure that all of its terms are clear and understood. This document, once signed, forms a contract in which the applicant agrees to abide by the terms and conditions outlined in its provisions. It is essential that all parties know and understand what the document says and how it impacts the green construction project.

GLOSSARY

The world of building codes is loaded with special terminology and acronyms. In fact, those who enforce and work with building codes have a language all their own. This glossary is provided in an effort to assist the reader in learning and speaking this important "language."

A

admixture a product such as a pozzolan, calcium chloride, or pigment that is sometimes added to concrete to improve or otherwise alter its properties such as workability, color, or curing temperature

aggregate coarse gravel and sand used in the preparation of concrete

as-built drawing a drawing and/or electronic file that identifies the specific, actual locations, configurations, and depths of installed utilities such as gas and water mains

B

backflow prevention device a one-way valve to be used between a potable water supply and some other water system, such as lawn sprinklers. Such a device allows potable water to flow through the service piping and into the irrigation or other system, but will not allow water to flow in the other direction to prevent contamination

backsiphonage occurs when used or contaminated water and other liquids are inadvertently drawn back into the potable water supply due to negative pressure in the system piping

basic wind speed minimum wind load used for the design of a dwelling that varies based on both geographic location and topographic conditions adjacent to the construction site

brazed a process similar to soldering, but with a higher temperature metal filler material

British thermal unit the quantity of heat needed to raise the temperature of 1 pound of water 1 degree Fahrenheit at a particular temperature (39° F)

building drain that portion of the drain piping contained within the building to a point 30 inches (762 mm) outside the dwelling

building envelope insulated walls, roofs, and floors surrounding the conditioned space within a dwelling; includes fenestrations (also see *fenestration*)

building official the official charged with the responsibility for administering the building, plumbing, electrical, mechanical, and other related codes

C

cleanout an intentional opening at prescribed locations in a plumbing drainage, waste, and vent system designed to facilitate access for the removal of obstructions and for general service and maintenance

code enhancement the addition of a certain safety feature such as a fire sprinkler system or fire-resistive construction in a dwelling that would not otherwise be required in exchange for the allowance of an alternative design, material, or method (also see *AMM* in the Acronym Index)

combustible material a material that will ignite and burn when exposed to heat or flame

combustion air a supply of air used in the combustion of a fuel

compressive strength the strength of a structural component and/or construction material when subjected to a compressive force (a force that tends to crush the component or press on the material)

concrete masonry unit   modular, blocklike units made of concrete that are used in masonry construction

conditioned spaces   spaces that are supplied with heated or cooled air that are designed to maintain a temperature of 68° F (20° C)

conflagration   the spread of a fire from one building to another

cross-connection   an inadvertent connection between two otherwise separate piping systems

cross-contamination   occurs when a potable water system is cross-connected to a nonpotable system and allows potentially contaminated water to mix with the potable water

D

daylight basement   a basement with at least one wall fully above grade

dead load   the actual weights of the construction materials used in the building; includes but is not limited to brick, wood framing, roof coverings, insulation, and gypsum board

deflection   actual amount of movement of a structural member under given load

duct   rigid metal or flexible plastic passageways used to convey air throughout a dwelling

E

ecoroof   a roof covered with various forms of vegetation

electrical service   the conductors and equipment for delivering energy from the serving utility to the wiring system of the premises served

engineered wood product   a manufactured wood product used for construction that has been tested and evaluated for acceptance under the code

environmental loads   similar to dead loads but caused by environmental conditions such as ice or snow on a roof

excavation   digging into the earth in the preparation of a site for the construction of a dwelling and related systems

exterior envelope   see *building envelope*

F

fenestration   an opening in the building envelope such as a window or a door

field change   a change during construction due to an unforeseen circumstance or due to a change in preference

fire resistive   an assembly used in construction that utilizes materials that offer a degree of fire protection and that have been tested in a laboratory setting and proven to be resistant to fire for a specified period of time

flame spread index   the rate at which flame burns along the surface of a material as determined during the ASTM E84 tunnel test

G

grade stamp   an indelible ink stamp applied to the surface of a piece of lumber or other wood product such as plywood, OSB, or particleboard that identifies its grade, species, and other information

grading   earthwork required to prepare a site for construction

granular fill   crushed rock, gravel, and similar used to change the profile of a construction site or as backfill behind foundation and retaining walls

gray water   water discharged from lavatories, bathtubs, showers, clothes washer, and laundry trays

grounded   connected to ground (earth) or to a conductive body that extends the ground connection

guard   a wall, railing, or similar element used to prevent falls from stairs, decks, elevated floors, and landings (also called a guardrail)

gypsum board   a paper-faced gypsum sheet used to form wall surfaces

H

headlap   wood shingle overlap

heat island effect   the tendency for urban areas to retain heat from the sun during the day and then release it at night, creating warm zones and increasing average ambient temperatures

hydropower   power generated by moving water

hydrostatic test   a test using liquid water under pressure

I

impervious surface a surface that does not absorb water

incised scored to facilitate the maximum absorption of a wood preservative into wood fibers under pressure

inspector a person who inspects construction at various phases to ensure compliance with building, electrical, plumbing and mechanical, and other codes

insulating concrete forms a modular concrete forming system made of foam plastic or other approved materials that is designed to remain in place after concrete is placed inside the forms; provides a layer of insulation around cast-in-place concrete

K

kip a unit of measure used in engineering equal to 1 kilopound or 1,000 pounds

L

lamp a light bulb

listed assembly a wall, roof, or floor configuration recognized by an approved standards body as having fire-resistive properties

live loads fluctuating loads induced on structural members within a dwelling or accessory structure from people and the contents placed within a dwelling

live/work unit a building that serves as both a place of residence and a business location

luminaire a lighting fixture

M

means of egress the exiting system, including emergency escape and rescue openings, door, or stairs

membrane penetration a pipe, vent, wire, or other material that enters only one surface of a wall, ceiling, or floor and does not penetrate entirely through an assembly

monolithic foundation footings, foundations, and slabs poured as one unit rather than individually; includes footings and foundation walls poured together, slab/footing combinations, and other configurations

N

net-zero dwelling a dwelling that generates at least as much energy as it uses

O

opaque wall a solid portion of a wall containing no windows or doors; may include gypsum board, studs, insulation, structural sheathing, exterior finish materials, or any other wall construction materials and configurations

opening a window, door, or vent in a wall, floor, or roof

P

penetration construction elements such as pipes, ducts, and wires that enter or penetrate into or go through a wall, ceiling, or floor

plans examiner a person who reviews construction plans prior to the issuance of a permit to ensure the proposed work will be in compliance with building, electrical, plumbing and mechanical, and other codes

plenum a chamber or space used as part of the air distribution system

potable water water that is safe for drinking and household use

prescriptive construction building construction that conforms to the basic requirements of the International Residential Code utilizing conventional construction materials and methods; does not employ or utilize alternative materials, designs, or construction methods; does not require engineering to demonstrate compliance with the code

products of combustion carbon monoxide and other waste products generated during the process of combustion

projection a roof overhang, cantilever, or architectural feature that extends beyond the exterior wall

R

raceway a conduit used for electrical wiring

rebar reinforcing steel bars of various diameters used to reinforce concrete

receptacles convenience outlets

reinforcing steel see *rebar*

return air air from within a dwelling that is returned to the furnace for reconditioning

roof sheathing roof plywood, oriented strand board, particle board, or other approved sheathing materials attached to the roof framing to provide a structural diaphragm

S

scupper an opening in a parapet wall that allows water to drain from a roof

seismic of or related to an earthquake

service entrance the point of entry in an electrical service

smoke alarm a device used to detect smoke in a dwelling and provide an alarm to warn occupants of a fire (also called a smoke detector)

smoke density index the quantity and density of smoke produced when a material burns as determined during the ASTM E84 tunnel test

span tables tables provided in the IRC that identify safe spans for lumber used in floors, roof, and ceiling framing under given loading conditions

structural properties those characteristics inherent to particular structural members that determine its ability to receive, resist, and distribute a force; dependent on the species and grade of lumber or the type and grade of steel used

supply air heated or cooled air that is distributed through ducts to condition a space

surcharge loads other than the weight of level soil adjacent to a retaining or foundation wall such as hillside slopes and vehicle weights

surface drainage storm water runoff from a dwelling site

T

tensile strength the strength of a structural component and/or construction material when subjected to a tensile force (a force that tends to stretch or elongate the component or material)

termiticide a pesticide used specifically for control of termites

thermal equilibrium two or more systems that have reached the same temperature and no longer exchange thermal energy

thermodynamics the study of heat and energy

through penetration a pipe, vent, wire, or other material or element that penetrates entirely through an assembly

townhouse also called rowhouse; single-family dwelling units constructed in groups of three or more attached units in which each unit extends from the foundation to roof and with a yard or public way on two or more sides

turbidity cloudiness or murkiness due to sediments suspended in water

U

Ufer ground A type of electrical earth grounding method named after its developer, Herbert G. Ufer, which employs a ground wire bonded to the reinforcing steel (rebar) in a reinforced concrete foundation or a length of wire embedded in concrete in an unreinforced concrete foundation; synonymous with the term "concrete encased grounding electrode"

unbraced length the length of a stud between lateral supports

under-floor area a crawl space or other service area beneath the floor framing, usually located under the first or main floor; may also include unfinished basements in some cases

W

wall assembly all components of a wall including studs, sheathing, insulation, interior and exterior wall coverings, papers, etc.; may be rated fire resistive for different time periods or not rated

water-distribution backflow the flow of water or other liquids into the potable water supply system from other than the intended source

wind forces forces generated by the wind as it blows against the walls and/or roof of a structure

ACRONYM INDEX

The world of building codes is loaded with special terminology and acronyms. In fact, those who enforce and work with building codes have a language all their own. This acronym index is provided in an effort to assist the reader in learning and speaking this important "language."

A

AC	acceptance criteria
ACCA	Air Conditioning Contractors of America
ACI	American Concrete Institute
AFCI	arc-fault circuit interrupter
AF & PA	American Forest and Paper Association
AFUE	Annual Fuel Utilization Efficiency
AHA	American Hardboard Association
AISI	American Iron and Steel Institute
AMM	alternative materials and methods
ANSI	American National Standards Institute
APA	The Engineered Wood Association; formerly the American Plywood Association
ASHRAE	American Society of Heating, Refrigerating and Air Conditioning Engineers
ASME	American Society of Mechanical Engineers
ASTM	ASTM International; formerly the American Society for Testing and Materials
AWG	American Wire Gauge
AWWA	American Water Works Association

B

BOCA	Building Officials and Code Administrators
BTU	British Thermal Unit

C

C	Celsius
CBO	certified building official
CMU	concrete masonry unit
COFS/Truss	Standard for Cold-Formed Steel Framing—Truss Design
CPVC	chlorinated polyvinyl chloride
CSSB	Cedar Shake and Shingle Bureau

D

DOC	Department of Commerce, United States
DWV	drainage, waste, and vent

E

EG	evaluation guideline
EJL	end-jointed lumber
EPA	Environmental Protection Agency, United States
ER	evaluation reports

F

F	Fahrenheit
FS	flame spread

G

GFCI	ground-fault circuit interrupter
GHBG	Green Home Building Guidelines

H

HPVA	Hardwood, Plywood, and Veneer Association

HSPF	Heating Season Performance Factor
HVAC	heating, ventilation, and air conditioning

I

IAS	International Accreditation Service
IBC	International Building Code
ICBO	International Conference of Building Officials
ICC	International Code Council
ICC-ES	International Code Council Evaluation Service
ICF	insulating concrete forms
IEBC	International Existing Building Code
IFGC	International Fuel Gas Code
ILAC	International Laboratory Accreditation Cooperation
IMC	International Mechanical Code
IPC	International Plumbing Code
IPSDC	International Private Sewage Disposal Code
IRC	International Residential Code

K

kg	kilograms
kPa	kilopascals

L

LED	light emitting diodes
LEED	Leadership in Energy and Environmental Design
LVL	laminated veneer lumber

M

MIT	Massachusetts Institute of Technology
mm	millimeters
MRA	Mutual Recognition Arrangement

N

NAHB	National Association of Home Builders
NAIMA	North American Insulation Manufacturers Association
NDS	National Design Standard

NEC	National Electrical Code
NES	National Evaluation Service
NFPA	National Fire Protection Association
NFRC	National Fenestration Rating Council
NGBS	National Green Building Standard
NSF	National Sanitation Foundation

O

OSB	oriented strand board

P

PCA	Portland Cement Association
PEX	cross-linked polyethylene tubing
PMG	Plumbing, Mechanical, and Fuel Gas Listing Program
PP	polypropylene plastic
PSF	pounds per square foot
PSI	pounds per square inch
PV	photovoltaic

Q

QC	quality control

S

SAVE	Sustainable Attributes Verification and Evaluation (Program)
SBCCI	Southern Building Code Congress International
SD	smoke developed index
SEER	Seasonal Energy Efficiency Ratio
SHGC	Solar Heat Gain Coefficient
SIP	structural insulated panel
SMACNA	Sheet Metal and Air Conditioning Contractors National Association

U

UL	Underwriters Laboratories
USGBC	United States Green Building Council

V

VAR	Verification of Attributes Report
VOC	volatile organic compounds

INDEX

Note: Page numbers followed by f refer to figures; page numbers followed by t refer to tables